ORGANIC SOLID-STATE REACTIONS

Organic Solid-State Reactions

Edited by

Fumio Toda
Okayama University of Science, Okayama, Japan

KLUWER ACADEMIC PUBLISHERS
DORDRECHT / BOSTON / LONDON

Library of Congress Cataloging-in-Publication Data is available

ISBN 1-4020-0227-0

Published by Kluwer Academic Publishers
PO Box 17, 3300 AA Dordrecht, The Netherlands

Sold and distributed in North, Central and South America
by Kluwer Academic Publishers,
101 Philip Drive, Norwell, MA 02061, USA

In all other countries, sold and distributed
by Kluwer Academic Publishers,
PO Box 322, 3300 AH Dordrecht, The Netherlands

Printed on acid-free paper

Printed and bound in Great Britain by Antony Rowe Limited

Contents

Preface

It has long been customary to carry out organic reactions in an organic solvent. Recently, however, it was discovered that most organic reactions proceed even in the solid state. Surprisingly, in many cases organic solid-state reactions proceed more efficiently and selectively than do solution reactions. This finding is important in relation to molecular dynamics in the solid state. It has also long been believed that molecules in crystal do not move freely. By mixing powdered reactant and reagent, however, the reaction occurs in the solid state. This shows that molecules move quite easily when the conditions are appropriate. For example, inclusion complex crystals are formed by mixing powdered host and guest compounds. These molecular movements can be monitored by the measurement of IR and UV spectra in the solid state. When chirality changes during molecular movement, it can also be monitored by CD measurement in the solid state. Furthermore, solid-state reactions, namely solvent-free reactions, are important from the viewpoint of green and sustainable chemistry. Both thermal and photochemical reactions can be carried out in the solid state. In this book are included three chapters on thermal reactions and three on photochemical reactions written by distinguished scientists in these fields. Such a book has never been published before.

Fumio Toda
Okayama, June 2001

Thermal Organic Reaction in the Solid State

Koichi Tanaka and Fumio Toda

Department of Applied Chemistry, Faculty of Engineering,
Ehime University Matsuyama, Ehime 790-8577, Japan
Department of Chemistry, Faculty of Science, Okayama University of Science
Okayama, Okayama 700-0005, Japan

1. Introduction

Solid-state organic reactions occur more efficiently and more selectively than in solution, since molecules in a crystal are arranged tightly and regularly. When greater selectivity is required in the solid-state reaction, host-guest chemistry techniques can be applied efficaciously. Reaction in the solid state of the guest compound as its inclusion complex crystal with a chiral host can give an optically active reaction product.

The occurrence of efficient solid-state reactions shows that the molecules reacting are able to move freely in the solid state. In fact, host-guest inclusion complexation can occur by simply mixing and grinding both crystals in the solid state. Surprisingly, solid-state complexation even occurs enantioselectively. These solid-state reactions can be easily monitored by measurement of infrared (IR), ultraviolet (UV) and circular dichroism (CD) spectra in the solid state.

In some cases, solid-solid reactions occur very efficiently and selectively in a water suspension medium. The solid-state reaction has many advantages: waste minimization, simple operation and easier product work-up. These factors are especially important in industry. In this chapter, thermal organic reactions in the solid state and in a water suspension medium are described.

2. Molecular Movement in the Solid State

It was found that inclusion complexation occurs efficiently by simple mixing and grinding of powdered host and guest compounds. For example, when an IR spectrum of a mixture of powdered 1,1,6,6-tetraphenylhexa-2,4-diyne-1,6-diol (**1**) and an equimolar amount of powdered benzophenone (**2**) was measured as a Nujol mull, it proved to be identical to that of their 1:1 inclusion complex (**3**) prepared by recrystallization of the two components from solution [1]. This result shows that formation of the complex by solid-state reaction occurs very rapidly.

F. Toda (ed.), Organic Solid-State Reactions, 1–46.
© 2002 *Kluwer Academic Publishers. Printed in Great Britain.*

The solid-state reactions can easily be monitored by measurement of the IR or UV spectrum as a Nujol mull. For example, formation of racemic 2,2'-dihydroxy-1, 1'-binaphthyl (**4**) by mixing and grinding of powdered (–)- (**4**) and (+)-enantiomers (**4**) in 1:1 ratio can be followed by successive IR measurements in Nujol mull (Figure 1) [2]. As the formation of **4** by solid-state reaction of (+)-**4** and (–)-**4** proceeds, the νOH absorptions of (+)-**4** and (–)-**4** at 3510 and 3435 cm^{-1} decrease and finally disappear within 1 h, and new νOH due to (±)-4 at 3490 and 3405 cm^{-1} appear. X-ray structures of (+)-**4** and (±)-**4** are shown in Figure 2. In (+)-**4** crystals, the molecules are stacked

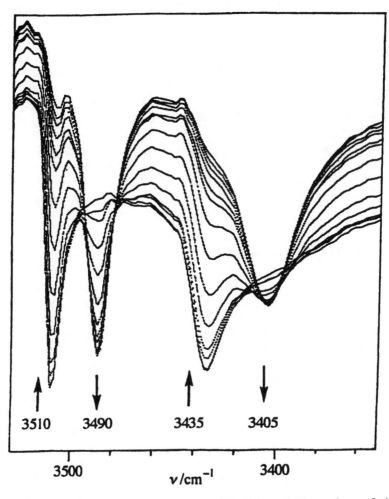

Figure 1 IR spectra of a 1:1 mixture of (+)-**4** and (–)-**4** in Nujol mull. Measured every 5 min for 1 h. Reprinted from *JCS Perkin Trans 2* (1997) 1880, by permission of the Royal Society of Chemistry.

(a)

(b)

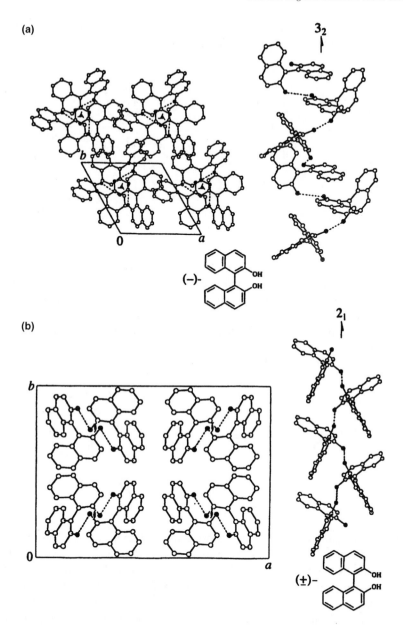

Figure 2 (a) View of the crystal structure of (+)-**4** along the *c*-axis and side view of the 3₂ helical structure. (b) View of the crystal structure of (±)-**4** along the *c*-axis and side view of the 2₁ helical structure. Reprinted from *JCS Perkin Trans 2* (1997) 1880, by permission of the Royal Society of Chemistry.

around a 3_2 screw axis to form a left-handed helical structure through a hydrogen bond. In (±)-**4** crystals, (+)-**4** and (−)-**4** molecules form a typical hydrogen-bonded helical structure along the 2_1 screw axis. The formation of (±)-**4** crystals from (+)-**4** and (−)-**4** crystals looks like a mutual penetration of 3_1 and 3_2 helical structures through the interface between enantiomeric crystals, followed by the conversion of three-fold screw structures into two-fold screw structures.

(R)-(+)-4	(S)-(−)-4	(±)-4
mp 211 °C	mp 211 °C	mp 219 °C
	nvOH: 3510, 3435 cm^{-1}	nvOH: 3490, 3405 cm^{-1}

When UV spectra of a 1:2 mixture of **1** and chalcone (**5**) were measured in the solid state every 10 min for 6 h, the absorption increased gradually as shown in Figure 3 [3]. As inclusion complexation proceeds, the number of complexed chalcone molecules (which has a co-planar structure) increases and hence the absorption coefficient increases. The co-planar structure of **6** in its complex with **1** has been proven by X-ray analysis [4].

Solid-state reaction occurs even enantioselectively. For example, inclusion complexation of (−)-**7b** and rac-**8** proceeds enantioselectively and gives a 1:1 complex of (−)-**7b** and (+)-**8**. After keeping a 1:2 mixture of powdered (−)-**7b** and powdered rac-**8** at room temperature for 2 days, uncomplexed (−)-**8** was extracted with hexane to give a 1:1 complex of (−)-**7b** and (+)-**8**, from which (+)-**8** of 88%ee was obtained in 32% yield by distillation in vacuo. From the hexane solution, (−)-**8** of 62%ee was isolated in 60% yield [5]. The most interesting application of the enantioselective solid-state reaction is the resolution of a racemic guest by distillation in the presence of a chiral host [6]. Heating of a 1:2 mixture of (−)-**7c** and rac-**9** at 70 °C/2 mm Hg gave (+)-**9** of 98%ee in 100% yield, and then further heating of the residue at 150 °C/2 mm Hg gave (−)-**9** of 100%ee in 98% yield. This resolution method is applicable to many kinds of rac-guest compounds such as alcohols, diols, epoxides, amino alcohols and cyclic amines.

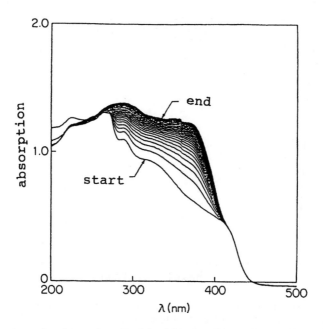

Figure 3 UV spectrum of a mixture of powdered **1** and **5** in the solid state (measured every 10 min for 6 h).

The enantioselective solid-state reaction can also be followed by measurement of CD spectra in Nujol mulls. The enantioselective inclusion complexation [7] of (−)-**7b** and (*S*)-(−)-pantolactone (**10**) was found to proceed in the solid state and the reaction was monitored by measurement of CD spectra as Nujol mulls. The CD spectrum of a mixture of powdered (−)-**7b** and two equivalents of (±)-**10c** in liquid paraffin was measured after 5 min of the preparation of the mull (Figure 4) [8]. The spectrum which shows

5

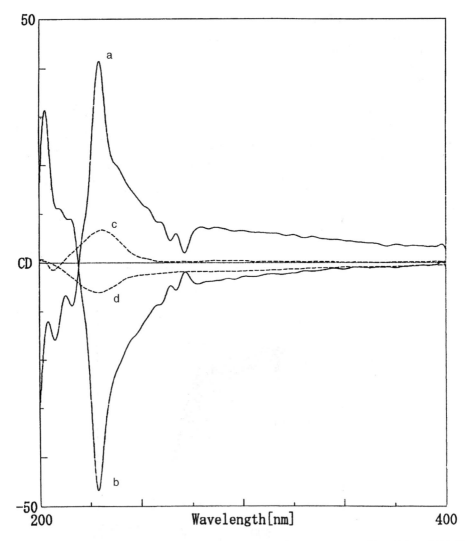

Figure 4   CD spectra of (a) (−)-**7b**·(+)-**10** and (b) (+)-**7b**·(−)-**10** complex prepared by mixing of (±)-**10** with (−)-**7b** and (+)-**7b**, respectively, and of (c) (−)-**10** and (d) (+)-**10** in Nujol mulls. Reprinted with permission from *Chem. Rev.* (2000) 1028. Copyright: American Chemical Society.

a (+)-Cotton effect is almost identical to that of an authentic inclusion crystal of (−)-**7b** and (−)-**10**. The assignment of the CD spectrum to the 1:1 inclusion complex of (−)-**7b** and (−)-**10** is reasonable, since **7b** itself shows only a very weak spectrum, and uncomplexed (+)-**10** left after the enantioselctive complexation also shows only a weak spectrum (Figure 4). On the other hand, a mixture of (+)-**7b** and two equivalents of (±)-**10** showed a CD spectrum with a (−)-Cotton effect by formation of the 1:1 inclusion complex of (+)-**7b** and (+)-**10** (Figure 4).

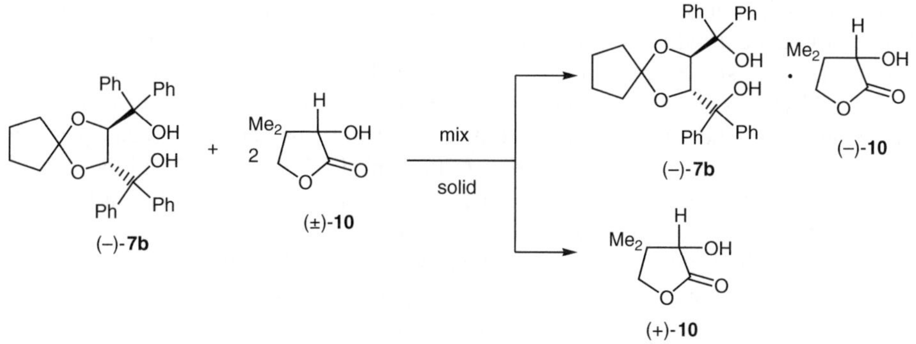

3. Thermal Organic Reaction in the Solid State

3.1. Oxidation

Some Baeyer–Villiger oxidations of ketones with *m*-chloroperbenzoic acid proceed much faster in the solid state than in solution. When a mixture of powdered ketone and 2 mol. equiv. of *m*-chloroperbenzoic acid was kept at room temperature, the oxidation product was obtained in the yield shown in Table 1 [9]. Each yield is higher than that obtained by reaction in CHCl$_3$ (Table 1).

Table 1 Yields of Baeyer–Villiger oxidation products in the solid state and in CHCl$_3$

Ketone	Reaction time	Product	Yield (%)	
			Solid state	CHCl$_3$
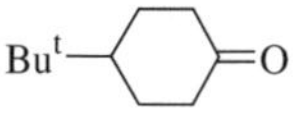 Bu^t—⬡=O	30 min	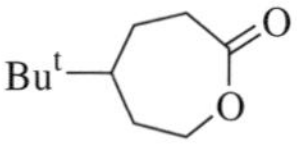 Bu^t—(ring)	95	94
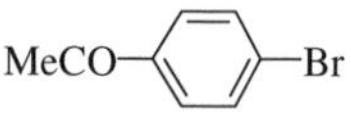 MeCO—⬡—Br	5 days	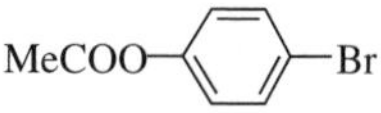 MeCOO—⬡—Br	64	50
PhCOCH$_2$Ph	24 h	PhCOOCH$_2$Ph	97	46
PhCOPh	24 h	PhCOOPh	85	13
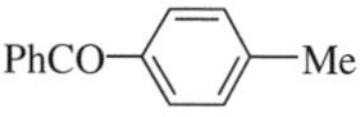 PhCO—⬡—Me	24 h	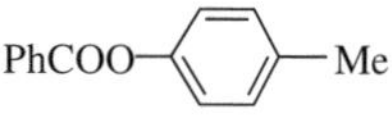 PhCOO—⬡—Me	50	12
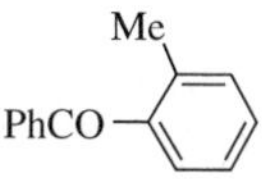 PhCO—(Me)⬡	4 days	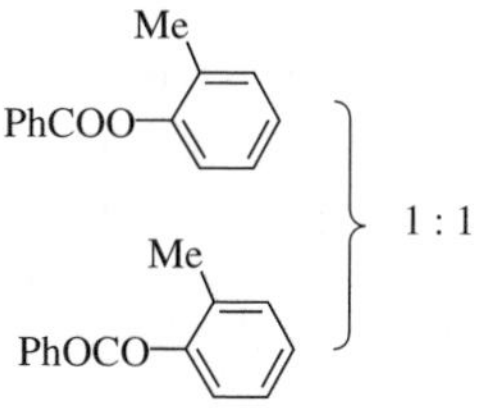 PhCOO—(Me)⬡ and PhOCO—(Me)⬡ 1:1	39	6

7

When a 1:1 inclusion complex of β-ionone **12** and optically active host compound (−)-**7c** was treated with two molar amounts of *m*-chloroperbenzoic acid in the solid state, a 1:1 inclusion complex of (−)-**7c** with (+)-**8** of 66% ee was obtained together with (−)-**13** of 72% ee [5]. This can be interpreted by an enantioselective inclusion complexation in the solid state between the initially formed (±)-**8** and (−)-**7c**, namely, (−)-**7c** includes (+)-**8** selectively in the solid state to form the 1:1 complex, and the uncomplexed (−)-8 is oxidized further to (−)-**13** with *m*-chloroperbenzoic acid. Similar solid-state kinetic resolution of dialkyl sulfoxides **14** was achieved by their enantioselective oxidation to dialkyl sulfones **15** with *m*-chloroperbenzoic acid in the presence of (−)-**11**. For example, a mixture of sulfoxide **14a** and (−)-**11** was kept at room temperature for 1 day, and then mixed with *m*-chloroperbenzoic acid and kept for a further 1 day. From the reaction mixture, (+)-**14a** of 37% ee was obtained in 38% yield (Table 2).

Table 2 Solid-state kinetic resolution of **14** by selective oxidation with MCPBA in the presence of (−)-**11**

Sulfoxide	Product		
	Yield (%)	[α]$_D$ (EtOH)	%ee
14a	(+)-**14a** 38	+44.2	37
14b	(+)-**14b** 51	+51.6	42.7
14c	(+)-**14c** 40	+25.6	25
14d	(+)-**14d** 7	+69	100

3.2. Reduction

Reduction of ketones with NaBH$_4$ also proceeds in the solid state. When a mixture of the powdered ketones and a ten-fold molar amount of NaBH$_4$ was kept in a dry box at room temperature with occasional mixing and grinding using an agate mortar and pestle for 5 days, the corresponding alcohols were obtained in the yields shown in Table 3 [10]. The reduction of benzophenones was studied in detail by *in situ* X-ray powder diffractometry and by ¹H NMR spectroscopy. The reaction efficiency was found to show a

Table 3 Reduction of ketones in the solid state by NaBH₄

Ketone	Alcohol	Yield (%)
Ph₂CO	Ph₂CH–OH	100
trans-PhCH=CHCOPh	trans-PhCH=CHCHPh OH PhCH₂CH₂CHPh OH } 1:1	100
![naphthyl]COMe	![naphthyl]CHMe OH	53
PhCHCOPh OH	meso-PhCH–CHPh OH OH	62
PhCH₂COPh	PhCH₂–CHPh OH	63
Buᵗ–⬡=O	Buᵗ–⬡–OH	92

strong dependence on the water content of the mixture and on the type of benzophenone [11]. Under strict water exclusion, the reaction does not occur whereas the presence of a small amount of water promotes the reaction.

In addition, enantioselective reduction of ketones is also found to proceed efficiently. For example, treatment of inclusion compounds of the ketones **16** in optically active host compounds with a BH₃-ethylenediamine complex in the solid state gave the chiral alcohols (*R*)-(+)-**17** of 20–60% ee as shown in Table 4 [12]. With the inclusion compounds of ketones with chiral host compounds, the BH₃-ethylenediamine complex would attack the ketones from the direction which gives the (*R*)-alcohols selectively.

$$
\underset{\textbf{16}}{\overset{O}{\overset{\|}{Ar-C-Me}}} \quad \xrightarrow[\text{Chiral host}]{2\ BH_3-NH_2CH_2CH_2NH_2} \quad \underset{\textbf{17}}{(R)\text{-}(+)\text{-}\ \overset{H}{\overset{|}{Ar\blacktriangleright C-Me}}\underset{OH}{}}
$$

Table 4 Yield, optical purity and absolute configuration of the alcohol obtained by the solid-state reduction

Host	Ar	Alcohol **17**		
		Yield (%)	% ee	Absolute configuration
(–)-**11**	Ph	96	44	R
(–)-**11**	*o*-tolyl	57	59	R
(–)-**11**	1-naphthyl	20	22	R
(–)-**7a**	1-naphthyl	32	22	R

9

Treatment of a 1:1 inclusion complex [13] of (*R*)-**18a** and (*R,R*)-(–)-**7a** with NaBH$_4$ in the solid state for 3 days gave (*R,R*)-(–)-**19a** of 100% ee in 54% yield [10]. The corresponding reaction of a 1:1 complex of (*S*)-**18a** and (*S,S*)-(1)-**7a** gave (*S,S*)-(1)-**19a** of 100% ee. The enone moiety of (*R*)-**18a** is masked by forming a hydrogen bond with the hydroxyl group of (*R,R*)-(–)-**7a** [14], so that the other carbonyl group is reduced selectively. Similar reduction of a 1:1 complex of (*R*)-**18b** and (*R,R*)-(–)-**7a** with NaBH$_4$ in the solid state gave (*S,S*)-(1)-**19b** of 100%ee in 55% yield.

3.3. Bromination reaction

The addition reaction of bromine to olefin is still difficult to control in most cases. For example, the reaction of trans-stilbene (**20**) with bromine in CH$_2$Cl$_2$ gives a 84:16 mixture of *meso*-**21** and *rac*-**22**. Treatment of crystalline **20** with bromine vapor gives *meso*-**21** and *rac*-**22** in a 62:38 mixture in only 20% yield. Very interestingly, however, when the reaction is carried out in the solid state by mixing of **20** and a solid brominating reagent (**25**), only *meso*-**21** is obtained in 71% yield (Table 5) [15]. Bromination of chalcone (**5**) in the solid state is also found to give only *erythro*-**23**, selectively (Table 6).

Table 5 Bromination of trans-stilbene (**20**) in solution, gas-solid and solid-solid media

Condition	Time (h)	Yield (%)	*meso*-**21**:*rac*-**22**
CH₂Cl₂	12	98	84:16
Gas-solid	64	20	62:38
Solid-solid	168	71	100:0

Table 6 Bromination of chalcone (**5**) in solution and solid-solid media

Condition	Time (h)	Yield (%)	*erythro*-**23**:*threo*-**24**
CH₂Cl₂	0.5	98	84:16
Solid-solid	4	89	100:0

Reaction of trans-o-stilbenecarboxylic acid **26** with bromine in solution gives trans-4-bromo-3-phenyl-3,4-dihydroisocoumarin **27** as the major product. On the other hand, treatment of powdered **26** with bromine vapor or with powdered **25** in the solid state at room temperature gave erythro-1,2-dibromo-1,2-dihydro-o-stilbenecarboxylic acid 28 selectively [16].

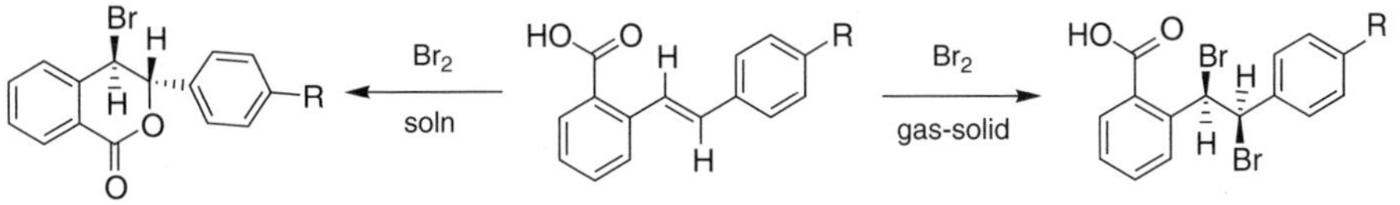

	26	a: R = H
27		b: R = Me
	28	c: R = Cl

3.4. Michael addition

Highly enantioselective Michael addition reactions of ArSH to 2-cyclohexenone have been achieved by using the optically active host compound (−)-**7c** [17]. For example, when a mixture of the powdered 1:1 inclusion complex of 2-cyclohexenone with (−)-**7c**, 2-mercaptopyridine and a catalytic amount of benzyltrimethylammonium hydroxide was irradiated with ultrasound for 1 h at room temperature, then (+)-**29a** of 80% ee was obtained in 51% yield (Table 7). Michael addition of thiols to 3-methyl-3-buten-2-one (**30**) in its inclusion crystal with (−)-**7c** also occurred enantioselectively to give Michael addition products (**31**) (Table 8).

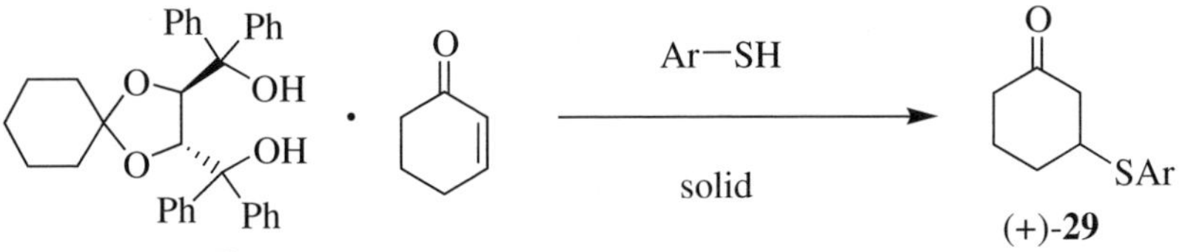

(R,R)-(−)-**7c**

(+)-**29**

Table 7 Enantioselective Michael addition of thiols to 2-cyclohexenone in its inclusion crystal

Ar	Reaction time (h)	Product	
		Yield	% ee
a	24	51	80
b	36	58	78
c	36	77	740

(R,R)-(−)-**7c**   **30**   Ar−SH   solid   (+)-**31**

Table 8 Enantioselective Michael addition of thiols to 3-methyl-3-buten-2- one in its inclusion crystal

Ar	Product	
	Yield	% ee
a	76	49
b	93	9
c	89	4
d	78	53

3.5. Elimination reaction

Dehydration reactions of alcohols also proceed efficiently in the solid state [18]. For example, when powdered 1,1-diphenylpropan-1-ol **32b** was kept in a desiccator filled with HCl gas for 5.5 h, pure 1,1-diphenylprop-1-ene **33b** was obtained in 99% yield. By the same method, **32a**, **32c**, and **32d** gave pure dehydration products, **33a**, **33c**, and **33d**, respectively in almost quantitative yields (Table 9).

$$\text{PhR}^1\text{C-CH}_2\text{R}^2 \quad \xrightarrow[\text{solid}]{\text{HCl gas}} \quad \text{PhR}^1\text{C=CHR}^2$$
$$\underset{\text{OH} \quad \textbf{32}}{|} \qquad\qquad\qquad \textbf{33}$$

Table 9 HCl-Catalysed dehydration of **32** in the solid state

	R^1	R^2	Reaction time (h)	Yield (%)
a	Ph	H	0.5	99
b	Ph	Me	5.5	99
c	Ph	Ph	8	100
d	o-Cl-C$_6$H$_4$	Me	4	97

Table 10 Cl$_3$CCO$_2$H-Catalysed dehydration of **32**

	R^1	R^2	Yield (%)	
			Solid	Benzene
a	Ph	H	99	–
b	Ph	Me	99	74
c	Ph	Ph	97	65

The dehydration reaction proceeds much faster by using Cl$_3$CCO$_2$H as a catalyst. For example, a mixture of powdered **32b** and an equimolar amount of Cl$_3$CCO$_2$H was kept at room temperature for 5 min, then the reaction mixture was washed with water and dried to give pure **33b** in 99% yield. However, the dehydration reaction in benzene gave **33** in relatively low yield (Table 10).

3.6. Cycloaddition reaction

The thermal crystal-to-crystal conversion of s-trans-1,1,6,6-tetraaryl-3,4-dibromo-1,2,4,5-hexatetraenes (**34a**, **34b** and **37b**) into the corresponding 3,4-bis(diarylmethylene)-1,2-dibromocyclobutenes (**36a**, **36b**, and **39b** and **40b**) via the s-cis-diallenes

13

(**35a**, **35b**, and **38b**) was found to proceed stereoselectively [19]. These thermal conversions involve two crystal-to-crystal reactions. First, the s-trans conformation of **34a** is rearranged to the s-cis conformer in the crystal to give **35a**. In the second step [2+2] conrotatory cyclization of **35a** occurs in the crystal to give **36a**. This cyclization proceeds stereoselectively; compound **34b** gave **36b**, and **37b** gave a 1:1 mixture of **39b** and **40b**, through a [2+2] conrotatory cyclization. For example, heating crystals of

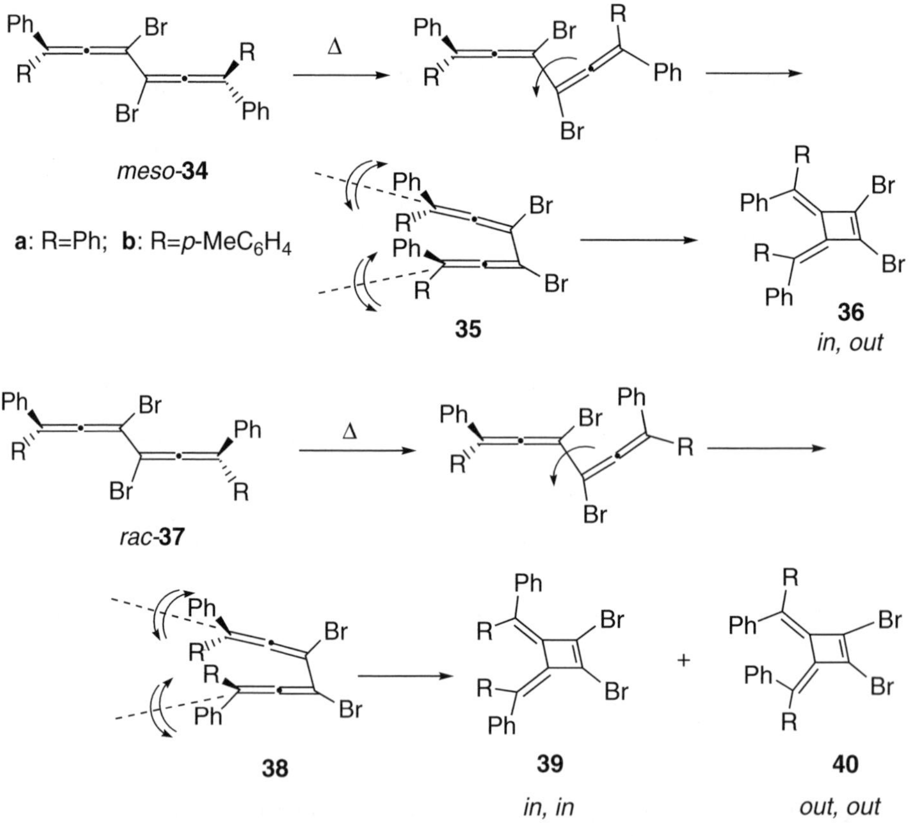

meso-**34**

a: R=Ph; b: R=*p*-MeC₆H₄

35

36
in, out

rac-**37**

38

39
in, in

40
out, out

34b at 135 °C gave the in, out isomer **36b**, while heating **37b** at 145 °C gave a 1:1 mixture of the in, in isomer **39b** and the out, out isomer **40b** in quantitative yields.

The differential scanning calorimetry (DSC) diagram of **37b** revealed a peak for an exothermic reaction at around 150 °C, which is attributable to the formation of a mixture of **39b** and **40b** and a peak for an endothermic conversion at around 193 °C, which is attributable to the melting point of the cyclization product (Figure 5). When the IR spectrum of a single crystal of **37b** was measured continuously every minute for 50 min at 125 °C, the signal at $\nu = 1927$ cm^{-1}(C=C=C) gradually decreased and finally

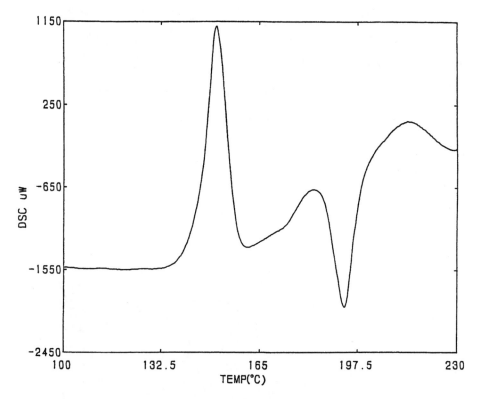

Figure 5 DSC diagram of rac-**37b**. Reprinted from *Angew. Chem. Int. Ed. Engl.* (1998) 2725, with permission from Wiley-VCH.

disappeared (Figure 6). This conversion from crystal-to-crystal was monitored through a microscope, and a molten state was not observed throughout the reaction, although the reaction product was no longer a single crystal (Figure 7). It is surprising that the thermal rearrangement and stereoselective cyclization occur so readily despite the required movement of a sterically bulky group in the crystal.

Heating of tetraallene derivatives (**41a–d**) at 140–200 °C in the solid state gave anthrodicyclobutene derivatives (**42a–d**), quantitatively [20]. For example, when **41a** was recrystallized from THF-toluene, a 1:2 inclusion crystal of **41a** and toluene was obtained as colorless prisms. Heating the colorless crystals of the 1:2 inclusion crystal at 180 °C on a hot plate for 30 min gave the yellow crystals of anthrodicyclobutene derivative (**42a**) in quantitative yield. The DSC measurement of the inclusion crystal showed a peak for an endothermic process at around 106 °C, which is attributable to the release of toluene from the complex, and a broad peak for an exothermic process at 189 °C, which is attributable to the thermal cyclization of **41a** to **42a**. The conversion from crystal-to-crystal was monitored through a microscope (Figure 8). Similarly, **42b–d** were obtained in quantitative yields upon heating of **41b–d** on a hot plate.

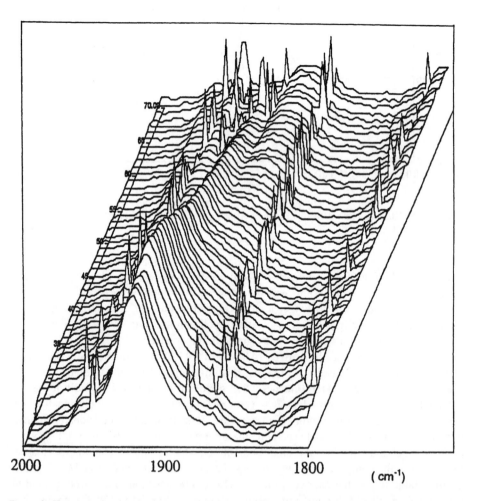

41

a: Ar = Ph

b: Ar = ⟨⟩-Me

c: Ar = ⟨⟩-F

d: Ar = ⟨⟩-Cl

42

2000   1900   1800

(cm⁻¹)

Figure 6 IR spectra showing the thermal reaction of rac-**37b** in the crystalline state at 125 °C. The spectrum was measured every minute for 50 min. Reprinted from *Angew. Chem. Int. Ed. Engl.* (1998) 2725, with permission from Wiley-VCH.

Figure 7 The thermal reaction of a crystal of rac-**37b** to a crystal of a mixture of **39b** and **40b** as observed through a microscope. The photos show the crystal before heating (a), as well as after 15 (b), 40 (c), and 80 min at 135 °C (d). Reprinted from *Angew. Chem. Int. Ed. Engl.* (1998) 2725, with permission from Wiley-VCH.

Heating of a colorless crystal of rac-**43** for 2 h at 140–150 °C affords green crystals of benzodicyclobutadiene **46** in quantitative yield (Figure 9) [21]. The DSC diagram of rac-**43** showed a peak for an exothermic reaction at around 160 °C, which is attributable to the formation of **46**. This reaction proceeds as follows: s-trans conformation of **43** is rearranged to the s-cis conformer (**44**), which gives **46** through [2+2] conrotatory cyclization via cyclobutene (**45**). In contrast, meso-**47** melts at 115–120 °C upon heating on a hot plate and then solidifies to give **46** via conrotatory cyclization of **48**.

rac-**43**   →   Δ   solid   →   **44**   →   Δ

46   ←   Δ   ←   **45**

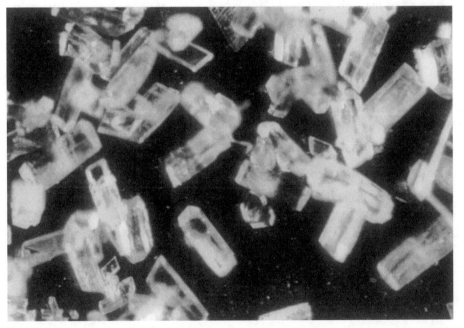

Figure 8 The thermal reaction of a crystal of **41a** to a crystal of **42a** as observed through a microscope. The photos show the crystal before heating (a) and after 30 min at 180 °C (b).

18

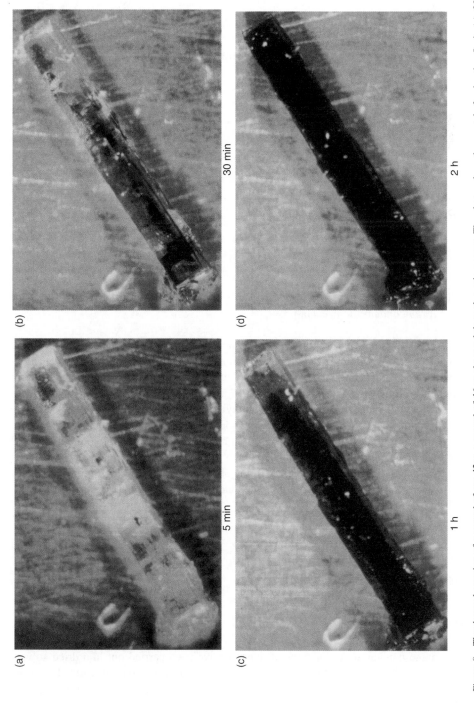

Figure 9 The thermal reaction of a crystal of rac-**43** to a crystal of **46** as observed through a microscope. The photos show the crystal after heating 5 min (a), 30 min (b), 1 h (c) and 2 h (d) at 140 °C.

3.7. Aldol condensation reaction

Some aldol condensation reactions proceed more efficiently and stereoselectively in the absence of a solvent than in solution [22]. When a slurry mixture of p-methylbenzalde-hyde (**50b**), acetophenone (**51b**), and NaOH was ground by pestle and mortar at room temperature for 5 min, the mixture turned to a pale yellow solid. The solid was combined with water and filtered to give p-methylchalcone (**53b**) in 97% yield. When the condensation was carried out in 50% aqueous EtOH according to the reported procedure for 5 min, the product was obtained only in 11% yield. The results of some other aldol reactions in the absence of a solvent are shown in Table 11. In most cases, the condensation reactions proceed more efficiently in the absence of a solvent than in 50% aqueous EtOH. Dehydration of the initially produced aldol **52** to chalcone **53** occurs more easily in the absence of a solvent.

3.8. Dieckmann condensation reaction

Dieckmann condensation reactions of diesters have been carried out in a dried solvent under high-dilution conditions in order to avoid intermolecular reaction. Recently, Dieckmann condensation reactions of diethyl adipate and pimelate were found to

Table 11 Aldol condensation reaction of **50** and **51** in the absence of solvent and in 50% aqueous EtOH

	50	51	Reaction time/min	Solvent	Yield (%)	
	Ar	Ar^t			52	53
a	Ph	Ph	30	–	10	0
				50%EtOH	0	36
b	*p*-MeC$_6$H$_4$	Ph	5	–	0	97
				50%EtOH	0	11
c	*p*-MeC$_6$H$_4$	*p*-MeC$_6$H$_4$	5	–	0	99
				50%EtOH	0	3
d	*p*-ClC$_6$H$_4$	Ph	5	–	0	98
				50%EtOH	18	59
e	*p*-ClC$_6$H$_4$	*p*-MeOC$_6$H$_4$	10	–	2	79
				50%EtOH	25	52
f	*p*-ClC$_6$H$_4$	*p*-BrC$_6$H$_4$	10	–	0	81
				50%EtOH	0	92
g	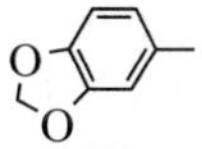 (benzodioxole group)	*p*-BrC$_6$H$_4$	10	–	0	91
				50%EtOH	0	0

proceed efficiently in the absence of a solvent, and the reaction products were collected by direct distillation of the reaction mixture [23]. For example, when diethyl adipate **54a** and Bu^{t}OK powder were mixed using a mortar and pestle for 10 min, the reaction mixture solidified. The solidified mixture was neutralized by addition of *p*-TsOH·H^{2}O and was distilled under 20 mm Hg to give **55a** in 82% yield. The solvent-free Dieckmann condensation of **54a** and **54b** also proceeds efficiently in the presence of powdered Bu^{t}ONa, EtOK and EtONa as summarized in Table 12.

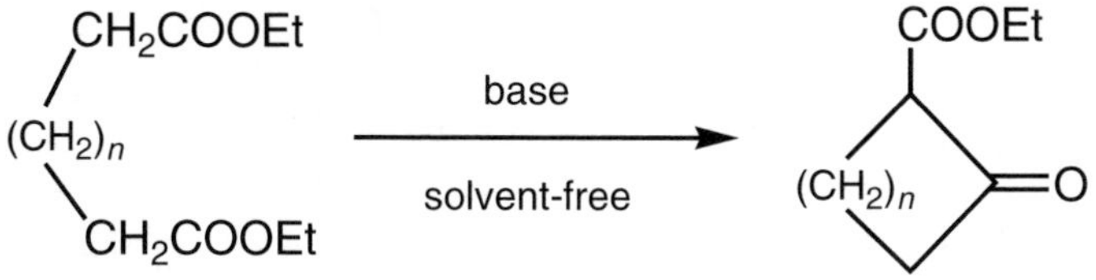

54 a: *n*=2
b: *n*=3

55

Table 12 Yield of solvent-free Dieckmann condensation reaction product **55a** and **55b**

Base	Yield of **55a**	Yield of **55b** (%)
Bu^{t}OK	82	69
Bu^{t}ONa	74	68
EtOK	63	56
EtONa	61	60

3.9. Stobbe condensation reaction

Stobbe condensation reaction also proceeds selectively in the solid state [24]. For example, Stobbe condensation reaction of cyclohexanone (56a) and diethyl succinate (57) in the presence of *t*-BuOK at room temperature and at 80 °C gave cyclohexylidenesuccinic acid (58a) and cyclohexenylsuccinic acid (59a), respectively. To an equivalent mixture of 56a and 57 was added powdered *t*-BuOK (1.2 eq.) in a mortar and well ground with a pestle at room temperature for 10 min. Then, the reaction mixture was neutralized with dilute HCl and the crystals formed were isolated by filtration to give cyclohexylidene-succinic acid (58a) in 75% yield. In contrast, when the reaction was carried out at 80 °C, cyclohexenylsuccinic acid (59a) was obtained in 92% yield. This method is very useful because both structural isomers (58a and 59a) can be prepared selectively simply by changing the reaction temperature. Similar treatment of 56b with 57 in the presence of *t*-BuOK for 10 min at room temperature and at 80 °C gave 58b and 59b in 55 and 85% yields, respectively (Table 13).

Stobbe condensation reactions of alkyl phenyl ketones (60) and diethyl succinate (57) were also found to proceed more efficiently and more selectively than those in solution. For example, a mixture of 60a, 57 and *t*-BuOK was mixed and ground using a mortar and pestle for 10 min at room temperature. The reaction mixture was neutralized with dilute HCl and then the crystalline product was isolated by filtration to give only 61a (*E:Z*=10:90) in 93% yield. In contrast, the reaction under refluxing in *t*-BuOH gave a 90:10 mixture of 61a and 62a, and the *E:Z* ratio of 61a was 65:34 (Table 14) [25].

Table 13 Stobbe condensation of 56 and 57

Cyclohexanone	Solvent	Temp (°C)	Yield (%)	58:59[a]
1a	*t*-BuOH	reflux	84	27:73
1a	None	rt	75	100:0
1a	None	40	88	65:35
1a	None	60	81	10:90
1a	None	80	92	0:100
1b	None	rt	55	95:5
1b	None	80	85	0:100

[a] The ratios were determined by ¹H-NMR.

O
‖
Ph—C—R + [CO₂Et / CO₂Et] →(t-BuOK, 10 min)→ EtO₂C—C(=...)—CO₂H / Ph...R (**61** a: R=Me, b: R=Et) + EtO₂C—CH—CO₂H / Ph—C(=CH...R)H (**62** a: R=H, b: R=Me)

60 a: R=Me **57**
b: R=Et

Table 14 Stobbe condensation of **60** and **57**

Ketone	Solvent	Temp (°C)	Yield (%)	**61:62**[a]	E:Z ratio of **61**
6a	t-BuOH	reflux	71	90:10	65:35
6a	None	rt	93	100:0	10:90
6a	None	80	82	90:10	55:45
6a	None	120	82	85:15	55:45
6b	t-BuOH	reflux	18	90:10	25:75
6b	None	rt	59	90:10	10:90
6b	None	80	87	35:65	45:55
6b	None	120	79	30:70	45:55

[a] The ratios were determined by ^{1}H-NMR.

3.10. Grignard reaction

Grignard reactions also occur in the solid state and some give different results from those in solution. For example, the reaction of ketones in the solid state gives more reduction products rather than addition products [26]. Dried Grignard reagents are obtained as a white powder by evaporation of the solvent *in vacuo* from the Grignard reagent prepared by the usual method in solution. One mole of powdered benzophenone (**2**) and three moles of the powdered dried Grignard reagent (**63**) were well mixed using an agate mortar and pestle, and the mixture then kept at room temperature for 0.5 h to give *tert*-alcohols (**64**) together with benzhydrol (**65**) in the yields shown in Table 15.

Ph_2CO + $RMgX$ →(rt, solid)→ Ph_2RCOH + Ph_2CHOH

2   **63**   **64**   **65**

63
a: R = Me; X = I
b: R = Et; X = Br
c: R = i-Pr; X = Br
d: R = Ph; X = Br

64
a: R = Me
b: R = Et
c: R = i-Pr
d: R = Ph

Table 15 Products and yiedls of Grignard reactions in the solid and in solution

Grignard reagent	Product and Yield (%)			
	Solid state		Solution	
63a	–	–	64a (99)	–
63b	64b (30)	65 (31)	64b (80)	65 (20)
63c	64c (2)	65 (20)	64c (59)	65 (22)
63d	64d (59)	–	64d (94)	–

3.11. *Reformatsky and Luche reaction*

Treatment of the aromatic aldehydes **66a–e** with ethyl bromoacetate **67** and Zn-NH$_4$Cl gave the corresponding Reformatsky reaction products **68a–e** in the yields shown in Table 16 [27]. The yields of **68** obtained in the solvent-free reaction are much better than that obtained by the reaction in dry benzene-ether solution. The solid-state Reformatsky reaction, which does not require the use of an anhydrous solvent, is thus advantageous.

Synthesis of homoallylic alcohols by the Luche reaction can also be carried out efficiently in the absence of a solvent. Treatment of aldehydes **66** with 3-bromopropene **69** and Zn-NH$_4$Cl in the absence of a solvent gave the corresponding Luche reaction products **70** in the yields shown in Table 17 [27].

$$RCHO \quad + \quad BrCH_2COOEt \quad \xrightarrow[\text{NH}_4\text{Cl}]{\text{Zn}} \quad RCH(OH)CH_2COOEt$$

$$\textbf{66} \qquad\qquad \textbf{67} \qquad\qquad\qquad\qquad\qquad \textbf{68}$$

Table 16 Reaction time and yield of the product **68** in the Reformatsky reactions of **66** and **67** in the absence of solvent

66	R	Reaction time (h)	Yield (%) of 68
a	Ph	2	91
b	Br—⟨ ⟩—	3	94
c	(benzodioxole)	3	94
d	(biphenyl)	3	83
e	(naphthyl)	3	80

$$\underset{\textbf{66}}{\text{RCHO}} + \underset{\textbf{69}}{\text{BrCH}_2\text{CH}{=}\text{CH}_2} \xrightarrow[\text{NH}_4\text{Cl}]{\text{Zn}} \underset{\textbf{70}}{\text{RCH(OH)CH}_2\text{CH}{=}\text{CH}_2}$$

Table 17 Reaction time and yield of the product **70** in the Luche reaction of **66** and **69** in the absence of solvent

66	R	Reaction time (h)	Yield (%) of **70**
a	Ph	4	99
e		4	87
f	*n*-Pentyl	1	83
g	*trans*-CH₃CH=CH-	1	98

3.12. Wittig reaction

Wittig reactions in the solid state of the inclusion compound of 4-methyl- and 3,5-dimethyl-cyclohexanone **71** and **72** and an optically active host compound with (carbethoxymethlene)triphenylphosphorane **75** gave optically active 4-methyl-**73** and 3,5-dimethyl-1-(carbethoxymethylene)cyclohexane **74**, respectively [28]. For example, when a mixture of the finely powdered 1:1 inclusion compound of (−)-**7b** and 4-methyl-cyclohexanone **71a** and phosphorane **75** was kept at 70 °C, the Wittig reaction was completed within 4 h. To the reaction mixture was added ether-petroleum ether (1:1), and then the precipitated triphenylphosphine oxide was removed by filtration. The crude product left after evaporation of the solvent from the filtrate was distilled *in vacuo* to give (−)-4-methyl-1-(carbethoxymethylene)cyclohexane **73a** of 42% ee in 73% yield. (−)-4-Ethyl-1-(carbethoxymethylene)cyclohexane **73b** of 45% ee and (−)-3,5-dimethyl-1-(carbethoxymethylene)cyclohexane **74** of 57% ee were also obtained in 73% and 58% yield, respectively.

3.13. Ylid reaction

Treatment of chalcones (**5**), cyclohexanones (**79**) and imines (**121**) with trimethylsulfonium Iodide (**77** or **78**) and KOH in the solid state gives cyclopropanones (**76**), oxiranes (**80** and **81**), and aziridines (**85**), respectively, in good yields [29]. For example, when a mixture of powdered chalcone (**5**), trimethylsulfonium Iodide (**78**) and KOH was kept at room temperature for 3 h, trans-1-benzoyl-2-phenylcyclopropane (**76**) was obtained in 79% yield.

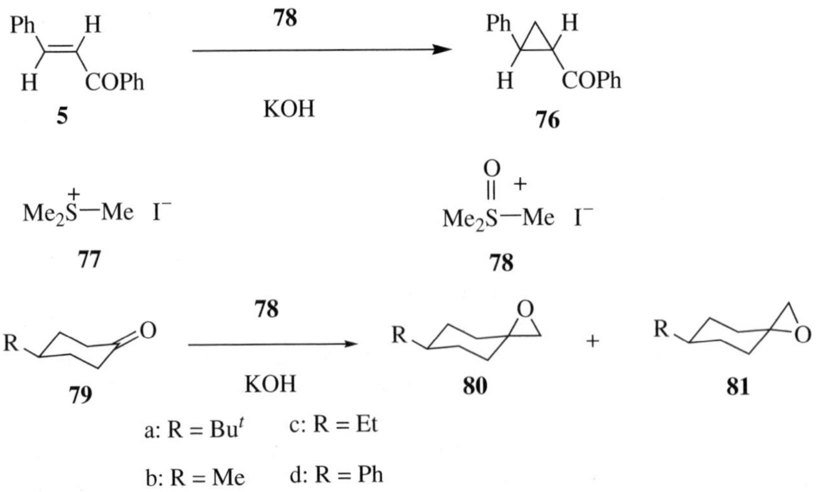

a: R = Bu^t c: R = Et

b: R = Me   d: R = Ph

Application of the same methylene transfer reaction in the solid state to the cyclohexanone derivatives **79**, gave the corresponding trans-oxiranes **80** selectively. For example, when a mixture of 4-tert-butylcyclohexanone **79a**, **78** and KOH was kept at room temperature for 3 h, the trans-isomer **80a** was obtained as the major product in 83% yield. In all solid-state methylene insertions using the reagent **78** the trans-isomer was the major product (Table 18).

Table 18 Yields of oxiranes **80** and **81** from the reaction of **79** with **78** and KOH at room temperature for 3 h in the solid state

Cyclohexanone 79	Yield (%) 80 + 81	Ratio 80:81
79a	83	97:3
79b	33	91:9
79c	75	92:8
79d	82	97:3

This reaction was improved to avoid use of any solvent throughout the reaction and the oxirane was isolated by a simple distillation technique [30]. For example, a mixture of propiophenone **82a, 77** and powdered t-BuOK was heated at 60 °C for 1 h in a flask, and then the reaction mixture was distilled using a Kugelrohr at 150 °C under 18 mm Hg, to give **83a** in 75% yield. By a similar procedure, **83b–f** were prepared from the corresponding ketones **82b–f** and the products isolated by distillation in the yields indicated in Table 19.

Table 19 Preparation of **83** by the combination of methylene transfer reaction to **82** in the absence of solvent and by Kugelrohr distilation

	Ketone		Reaction conditions		Product	
	Ar	R	Temp (°C)	Time (h)	Yield (%)	
82a	Ph	Et	60	1	**83a**	75
82b	Ph	i Pr	70	1	**83b**	89
82c	Ph	⬡—	70	10	**83c**	91
82d	p-MeC$_6$H$_4$	Et	70	2	**83d**	86
82e	P-BrC$_6$H$_4$	Me	70	5	**83e**	64
82f	2-Naphthyl	Me	70	3	**83f**	86

Methylene addition to the imines **84** in the solid state occurred at a relatively high temperature. For example, aziridines **85a–d** were obtained by the reaction of **84a–d** with **78** and KOH at 50 °C for 3 h in the solid state in 56, 34, 38, and 36% yields, respectively [29].

a: Ar = Ar' = Ph

b: Ar = Ph, Ar' = p-MeOC$_6$H$_4$

c: Ar = p-ClC$_6$H$_4$, Ar' = p-MeC$_6$H$_4$

d: Ar = p-MeC$_6$H$_4$, Ar' = p-MeOC$_6$H$_4$

Enantioselective methylene transfer also occurs in the solid state. By treatment of chalcone **5** with (+)-S-methyl-S-phenyl-N-(p-tolyl)sulfoximide **86** and Bu^{t}OK at room temperature, (+)-**76** of 24% ee was obtained in 94% yield [29].

3.14. Condensation reaction of amine and aldehyde

Condensation reactions of anilines **86** and aromatic aldehydes **87** to azomethines **88** were also found to proceed very efficiently in the absence of a solvent. Various kinds of azomethines **88** were obtained quantitatively (100% yield at 100% conversion) as hydrates by grinding together the solid anilines and solid benzaldehydes [31]. The reaction was determined to proceed without passing through liquid phases by measurements using atomic force microscopy (AFM).

Table 20 Yield of **90** and **91** produced by treatment of **89** with Zn-ZnCl$_2$ at room temperature for 3 h in 50% aqueous THF and in the solid state

89	Solvent	Yield (%)		meso:dl ratio in 91
		90	**91**	
a, X=H	aq. THF	39	11	50:50
	–	trace	46	60:40
b, X=Me	aq. THF	81	7	50:50
	–	trace	87	70:30
c, X=Cl	aq. THF	82	16	50:50
	–	25	65	80:20
d, X=Br	aq. THF	72	27	50:50
	–	19	55	70:30
e, X=Ph	aq. THF	49	38	80:20
	–	2	64	70:30

$$\text{ArCOAr'} \xrightarrow[\text{solid}]{\text{Zn-ZnCl}_2} \underset{\overset{|}{\text{OH}} \quad \overset{|}{\text{OH}}}{\text{ArAr'C}-\text{CArAr'}}$$

$$\textbf{92} \qquad\qquad\qquad\qquad\qquad \textbf{93}$$

$$\text{Ar–NH}_2 \quad + \quad \text{Ar'–CHO} \xrightarrow[\text{solid}]{\text{mix}} \text{Ar–N=CH–Ar'} \quad + \quad \text{H}_2\text{O}$$

$$\textbf{86} \qquad\qquad \textbf{87} \qquad\qquad\qquad\qquad \textbf{88}$$

Ar = 4-MeC$_6$H$_4$ Ar = 1-Naphthyl

Ar = 4-MeOC$_6$H$_4$ Ar' = 4-ClC$_6$H$_4$

Ar = 4-NO$_2$C$_6$H$_4$ Ar' = 4-BrC$_6$H$_4$

Ar = 4-ClC$_6$H$_4$ Ar = 4-NO$_2$C$_6$H$_4$

Ar = 4-BrC$_6$H$_4$ Ar = 4-HOC$_6$H$_4$

Ar = 4-HOC$_6$H$_4$ Ar = 4-HO, 3-MeOC$_6$H$_4$

Ar = 4-(4-H$_2$NC$_6$H$_4$)C$_6$H$_4$

3.15. Pinacol coupling reaction

The Zn-ZnCl$_2$ reagent is effective for the coupling of aromatic aldehydes and ketones to produce α-glycols in the solid state [32]. For example, when a mixture of **89a**, Zn, and ZnCl$_2$ was kept at room temperature for 3 h, the α-glycol **91a** was obtained in 46% yield. Similar treatment of benzaldehyde derivatives **89b–e** with the reagent gave mainly α-glycols **91b–e** in the yields shown in Table 20. The coupling reaction of aromatic ketones **92** with Zn-ZnCl$_2$ is more selective, and only the α-glycols **93** were produced (Table 21).

Table 21   Yield of **93** produced by treatment of **92** with Zn-ZnCl$_2$ in 50% aqueous THF and in the solid state

Ar	Ar'	Solvent	Reaction time (h)	Temp (°C)	Yield of **93** (%)
Ph	Ph	aq. THF	1	rt	84
		–	6	rt	86
p-MeC$_6$H$_4$	p-MeC$_6$H$_4$	aq. THF	3	rt	84
		–	84	70	30
p-ClC$_6$H$_4$	p-ClC$_6$H$_4$	aq. THF	1	rt	92
		–	6	70	39
Ph	p-ClC$_6$H$_4$	aq. THF	2	rt	90
		–	6	70	26
(fluorenone structure)		aq. THF	0.5	rt	94
		–	3	rt	92

3.16. Phenol coupling reaction

Some oxidative coupling reactions of phenols in the presence of $FeCl_3 \cdot 6H_2O$ proceed faster and more efficiently in the solid state than in solution. For example, when a mixture of β-naphthol **94** and $FeCl_3 \cdot 6H_2O$ was finely powdered using an agate mortar and pestle, and then the mixture was kept at 50 °C for 2 h, bis-β-naphthol **4** was obtained in 95% yield after decomposition of the reaction mixture with dilute HCl [33]. In contrast, heating a solution of **94** and $FeCl_3 \cdot 6H_2O$ in 50% aqueous MeOH under reflux for 2 h gave **4** in 60% yield. Some reactions are accelerated by irradiation with ultrasound. For example, when a mixture of finely powdered **95** and two molar amounts of $[Fe(DMF)_3Cl_2][FeCl_4]$ was irradiated with ultrasound at 50 °C for 9 h in the solid state, 9,9'-bisphenanthrol **96** was obtained in 68% yield. Conversely, holding of a solution of **95** and two molar amounts of $[Fe(DMF)_3Cl_2][FeCl_4]$ in CH_2Cl_2 at room temperature for 48 h gave **96** in only 33% yield in addition to byproducts such as 9-phenanthrone and 9,10-phenanthrenequinone. By a similar procedure, the bisphenol derivative **98** was obtained in 30% yield by oxidative coupling of **97** in the solid state [34].

3.17.   *Glaser coupling of acetylenic compound*

Glaser coupling of acetylenic compounds proceeds more efficiently in the solid state than in water [35]. When a mixture of powdered cuprous phenylacetylide **99a** and CuCl$_2$-2H$_2$O was kept at room temperature for 3 h, diphenyldiacetylene **100a** was obtained in 60% yield. By the same method, **99b–e** gave **100b–e** (Table 22).

$$Ar\text{-}C{\equiv}C\text{-}Cu \xrightarrow[\text{solid}]{CuCl_2\cdot 2H_2O} Ar\text{-}C{\equiv}C\text{-}C{\equiv}C\text{-}Ar$$

$$\mathbf{99} \qquad\qquad\qquad\qquad \mathbf{100}$$

Table 22   Glaser coupling reaction in the solid state and in water

	Ar	Yield (%)	
		Solid state	Water
a	Ph	60	40
b	p-MeC$_6$H$_4$	35	21
c	p-PhC$_6$H$_4$	67	–
d	2,3,5,6-(Me)$_4$-C$_6$H	42	25
e	PhOCH$_2$	74	–

Eglinton coupling reaction could also be applied to the reaction of propargyl alcohols in the solid state. When a mixture of the powdered propargyl alcohols **101** and CuCl$_2$-(pyridine)$_2$ complex was reacted under heating, the coupling products **102** were obtained in 50–70% yields [35].

Oxidative coupling reaction of (±)-**103** in pyridine gave the corresponding cyclic dimer (±)-**104**. When the reaction, however, was carried out in the solid state using Cu(OAc)$_2$-2Py complex, the linear coupling product **106** was obtained. On the other hand, (–)-**103** gave the optically active polymer **106** in solution and the optically active dimer (–)-**105** in the solid state as a major product, respectively [35].

$$2\ \underset{\underset{\mathbf{101}}{OH}}{RR'C\text{-}C{\equiv}CH} \xrightarrow[\text{solid}]{CuCl_2\cdot 2Pyridine} \underset{\underset{\mathbf{102}}{OH\qquad OH}}{RR'C\text{-}C{\equiv}C\text{-}C{\equiv}C\text{-}CRR'}$$

a: R=Ph; R'=Ph
b: R=Ph; R'=o-ClC$_6$H$_4$
c: R=p-MeC$_6$H$_4$; R'=p-MeC$_6$H$_4$
d: R=Ph; R'=2,4-(Me)$_2$-C$_6$H$_3$
e: R=2,4-(Me)$_2$-C$_6$H$_3$; R'=o-ClC$_6$H$_4$
f: R=Ph; R'=Me

3.18. Substitution reaction

Conversion of secondary and tertiary alcohols **107** into the corresponding chlorides **108** also proceeds efficiently when powdered **107** is exposed to HCl gas in a desiccator (Table 23) [18].

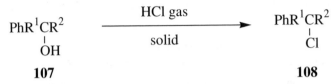

$$PhR^1CR^2 \xrightarrow[\text{solid}]{\text{HCl gas}} PhR^1CR^2$$

107 **108**

Table 23 S$_N$-reaction of **107** in the solid state

	R^1	R^2	Reaction time (h)	Yield (%)
a	Ph	H	5	92
b	Ph	Ph	1.5	97
c	o-Cl-C$_6$H$_4$	H	10	94

3.19. Etherification reaction

Formation of the ether **110** by treatment of alcohol **109** with TsOH in the solid state proceeds efficiently. For example, when a mixture of powdered 4-methylbenzhydrol **109e** and an equimolar amount of TsOH was kept at room temperature for 10 min, the corresponding ether **109e** was obtained in 96% yield. By the same procedure, other ethers were also synthesized in good yields (Table 24) [18]. In order to know why the etherification of **109** proceeds more efficiently in the solid state, the X-ray crystal structure of **109a** was analyzed. The data showed that two molecules of **109a** form a hydrogen-bonded dimer, in which the pair of **109a** molecules are located in close proximity.

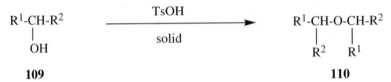

$$R^1\text{-CH-R}^2 \xrightarrow[\text{solid}]{\text{TsOH}} R^1\text{-CH-O-CH-R}^2$$

109 **110**

Table 24 TsOH-catalyzed etherification of **109** to **110** in the solid state and in solution

	109		Yield (%) of 110		
	R^1	R^2	Solid	Benzene	MeOH
a	Ph	Ph	95	45	34
b	Ph	o-ClC$_6$H$_4$	94	73	37
c	Ph	p-BrC$_6$H$_4$	98	53	50
d	Ph	p-NO$_2$C$_6$H$_4$	74	58	1
e	Ph	p-MeC$_6$H$_4$	96	52	10

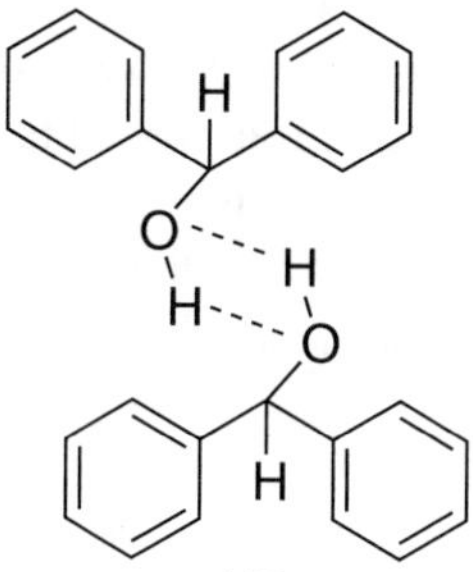

109a

3.20. Naphthopyran synthesis

p-TsOH catalyzed condensation reactions of 1,1-diaryl-2-propyn-1-ol (**111**) and 2-naphthol (**94**) in the solid state gave 3,3-diaryl-3H-naphtho[2,1-b]pyran (**115**) via Claisen rearrangement [36]. For example, a mixture of 1,1-diphenyl-2-propyn-1-ol (**111a**), 2-naphthol (**94**), p-TsOH (0.1 eq) and a small amount of silica-gel was well ground for 10 min at room temperature using a mortar and pestle and was kept for 1 h. The reaction mixture was chromatographed on silica gel to give 3,3-diphenyl-3H-naphtho[2,1-b] pyran (**115a**) in 56% yield.

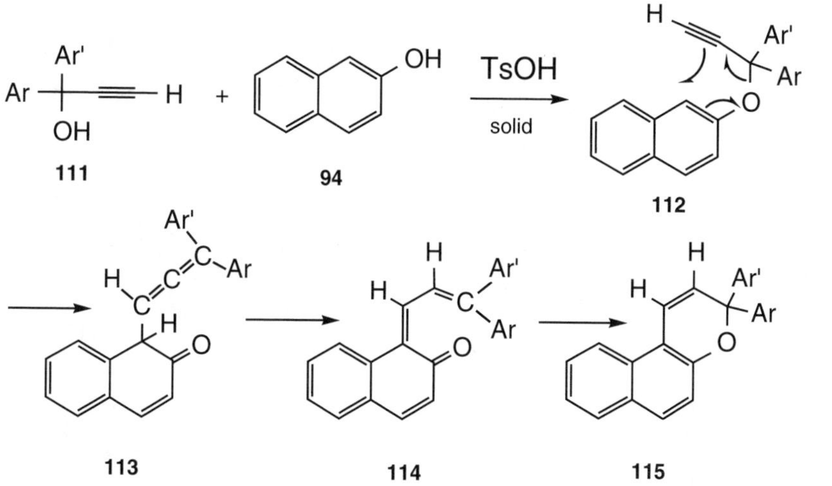

a: Ar=Ar'=Ph

b: Ar=Ar'=4-methyphenyl

c: Ar=Ar'=4-chlorophenyl

d: Ar=Ph; Ar'=4-methoxyphenyl

e: Ar=Ph; Ar'=2,4-dimethyphenyl

f: Ar.Ar'=fluorenyl

This solid-state reaction involves the following two steps. First 1,1-diphenyl-2-propyn-1-ol (**111a**) and 2-naphthol (**94**) are condensed to give propargyl ether (**112a**) in the presence of *p*-TsOH. In the second step, Claisen rearrangement of **112a** occurred to give **113a**, and finally naphthopynan **115a** was obtained via cyclization of quinodimethane (**114a**). Similar treatment of **111b–f** with **94** gave the corresponding naphthopyrans (**115b–f**) in the yields of 30–63% (Table 25).

Table 25 One-pot synthesis of naphthopyrans in the solid state

Compound	Ar	Ar'	Yield (%)	Mp (°C)
115a	Ph	Ph	56	162–167
115b	⟨—⟨benzene⟩—Me⟩	⟨—⟨benzene⟩—Me⟩	56	156–161
115c	⟨—⟨benzene⟩—Cl⟩	⟨—⟨benzene⟩—Cl⟩	30	188–192
115d	Ph	⟨—⟨benzene⟩—OMe⟩	58	154–157
115e	Ph	⟨—⟨benzene⟩—Me, Me⟩	63	122–124
115f	⟨biphenyl, 2,2'-dimethyl⟩		44	194–200

Naphthodipyran derivatives (**117** and **119**) were also prepared efficiently by the solid-solid reaction between **111a** and 2,6- (**116**) and 2,7-dihydroxynaphthalene (**118**). When the reaction of **94** with 1,1,6,6-tetraaryll-2,5-hexadiyn-1,6-diol (**1**) was carried out, bis-naphthopyran derivatives (**120**) were obtained in good yield.

3.21. *Coumarin synthesis*

Coumarin derivatives are important chemicals in the perfume, cosmetic, agricultural and pharmaceutical industries. However, the conventional methods for coumarin synthesis require drastic conditions. For example, 4-methyl-7-hydroxycoumarin has been prepared by stirring a mixture of resorcinol and ethyl acetoacetate in concentrated H_2SO_4 for 12–24 h [37]. Recently, a simple and efficient synthesis of coumarins via the Pechmann and Knoevenagel condensation reactions under solvent-free conditions has been developed [38].

To an equivalent mixture of resorcinol (**121a**) and ethyl acetoacetate (**122a**) was added TsOH in a mortar and ground well with a pestle at room temperature. The mixture was heated at 60 °C for 10 min under atmosphere. After cooling, water was added to the reaction mixture and the crystalline products were collected by filtration to give

7-hydroxy-4-methylcoumarin (**123a**) in 98% yield. Similarly, solvent-free Pechmann reactions of **121** and **122** afforded **123b**, **123c**, **123e**, **123f**, and **123g** in 92, 79, 69, 81, and 66% yields, respectively (Table 26). This method is very useful because **123b**, **123d**, **123e** and **123f** have not been obtained from the reaction in H_2SO_4; however, **123d** was not formed either in H_2SO_4 or in the absence of a solvent.

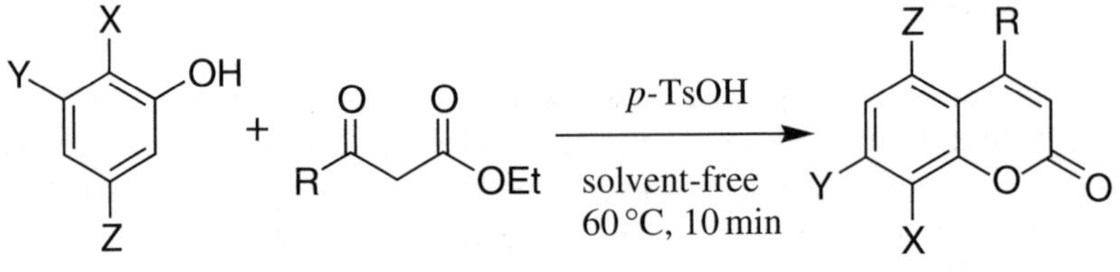

121 **122** **123**

a: X = H; Y = OH; Z = H a: R = Me
b: X = OH; Y = OH; Z = H b: R = Ph
c: X = H; Y = OH; Z = OH c: R = CH_2CO_2Et
d: X = H; Y = Me; Z = OH d: R = CH_2Cl

Table 26 Solvent-free Pechmann reactions of phenols (**121**) and β-keto esters (**122**)

123	X	Y	Z	R	Yield / %	
					Solvent-free	in H_2SO_4
a	H	OH	H	Me	98	82–90
b	H	OH	H	Ph	92	0
c	H	OH	H	CH_2CO_2Et	79	40
d	H	OH	H	CH_2Cl	0	0
e	OH	OH	H	Me	69	0
f	H	OH	OH	Me	81	0
g	H	Me	OH	Me	66	68

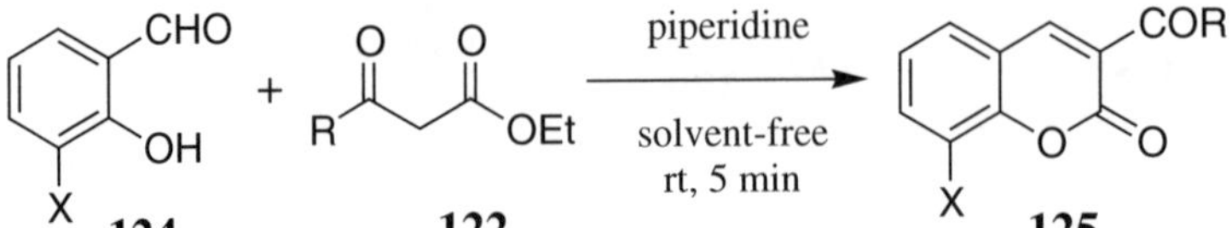

124 **122** **125**

a: X = H; a: R= Me
b: X = OMe; b: R= Ph
c: R= CH_2CO_2Et
d: R= OEt

Table 27 Solvent-free Knoevenagel reactions of salicylaldehyde (**124**) and β-keto esters (**122**)

125	X	R	Yield / %
a	H	Me	99
b	H	Ph	97
c	H	OEt	95
d	H	CH$_2$CO$_2$Et	73
e	OMe	Me	99
f	OMe	Ph	99
g	OMe	OEt	97
h	OMe	CH$_2$CO$_2$Et	91

Solvent-free Knoevenagel reactions of salicylaldehydes (**124**) and β-keto esters (**122**) were also found to proceed efficiently and under milder reaction conditions than in EtOH solution. For example, a mixture of salicylaldehyde (**124a**), diethyl malonate (**122e**) and a few drops of piperidine was ground well for 5 min at room temperature. The reaction mixture was neutralized with dilute HCl and then the crystalline product was isolated by filtration to give 3-carboethoxycoumarin (**125c**) in 95% yield. Similarly, substituted coumarin derivatives were obtained in high yields (Table 27).

It has been reported that the Knoevenagel reaction of 2-hydroxy-3-methoxy-benzaldehyde (**124b**) and ethyl cyanoacetate (**126**) affords 8-methoxy-2-oxo-2*H*-chromene-3-carbonitrile (**127**) *via* intramolecular cyclization of *Z*-**128** in 35% yield under reflux in EtOH [39]. Very interestingly, however, the condensation reaction of **124b** and **126** in the absence of a solvent gave 8-methoxy-2-oxo-2*H*-chromene-3-carboxylic acid ethyl ester (**125g**) in 65% yield along with a small amount of **127** (11% yield). Compound **125g** might be obtained via hydrolysis of iminolactone **130** formed by intramolecular cyclization of *E*-**129**.

3.22. Aminolysis

The solid-state reaction of a hydrazine inclusion complex with an ester gives a pure hydrazide [40]. Recrystallization of hydroquinone **131** from aq. hydrazine gives a 1:1 inclusion complex **132** of anhydrous hydrazine with **131** as colorless crystals. A mixture of powdered dimethyl terephthalate **133a** and **132** was kept under a nitrogen atmosphere at 100–125 °C for 25 h. To the reaction mixture was added MeOH and almost pure dihydrazide **134a** was obtained simply by filtration in 88% yield.

a: *para*
b: *meta*

3.23. Pinacol rearrangement

The pinacol rearrangement in the solid state was found to proceed faster and more selectively than that in solution [41]. When a 1:3 molar ratio of powdered **135** and p-TsOH was kept at 60 °C, the rearrangement products **136** and **137** were obtained in the yields shown in Table 28. The hydride migrates more easily than the phenyl anion in **135**, and the yield of **136** is higher than that of **137** in all reactions shown above. In contrast, when a mixture of powdered **135** and CCl_3CO_2H was kept at 20 °C for the period shown in Table 29, **137** was obtained as the major product. The reaction efficiency is dramatically enhanced when the water formed during the reaction is continuously removed under reduced pressure.

The mechanism of solid-state pinacol rearrangement of **138** to **139** has been studied by atomic force microscopy [42] and crystal structure analysis [43] of **138**.

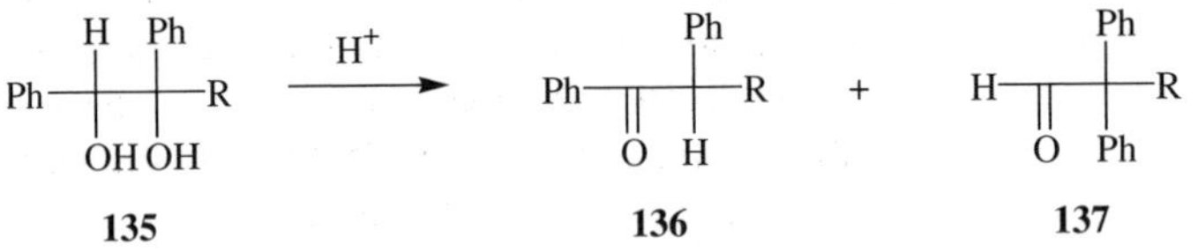

a: R = ph
b: R = o-MeC$_6$H$_4$
c: R = m-MeC$_6$H$_4$
d: R = p-MeC$_6$H$_4$
e: R = p-MeOC$_6$H$_4$
f: R = p-ClC$_6$H$_4$

Table 28 Pinacol rearrangement catalyzed by p-TsOH at 60 °C

Pinacol	Reaction time (h)	Yield (%)	
		136	**137**
135a	2.5	89	8
135b	0.5	45	29
135c	0.3	70	30
135d	0.7	39	19
135e	0.7	89	0
135f	1.0	54	41

Table 29 Pinacol rearrangement catalyzed by CCl_3CO_2H at 20 °C

Pinacol	Reaction time (h)	Yield (%)	
		136	**137**
135a	2.0	21	68
135b	3.0	38	62
135c	3.0	18	43
135d	2.5	38	62
135e	1.0	59	30
135f	2.0	30	64

3.24. Benzilic acid rearrangement

Benzilic acid rearrangement has long been carried out by heating benzil derivatives and alkali metal hydroxide in aqueous organic solvent. However, the benzilic acid rearrangements proceed more efficiently and faster in the solid state than in solution. For example, a mixture of finely powdered benzil **140a** and KOH was heated at 80 °C for 0.2 h, and the reaction product was mixed with 3N HCl to give benzilic acid **141a** as colorless needles. Similar treatment of benzil derivatives **140b–f** in the solid state also gave the corresponding benzilic acid **141b–f** (Table 30) [44].

Table 30 Yield of benzilic acid **141** produced by treatment of benzil **140** with KOH at 80 °C in the solid state

140	Ar	Ar′	Reaction time / h	Yield / % of 141
a	Ph	Ph	0.2	90
b	Ph	p-Cl-C$_6$H$_4$	0.5	92
c	p-Cl-C$_6$H$_4$	p-Cl-C$_6$H$_4$	6	68
d	Ph	p-NO$_2$-C$_6$H$_4$	0.1	93
e	m-NO$_2$-C$_6$H$_4$	m-NO$_2$-C$_6$H$_4$	0.1	72
f	Ph	p-MeO-C$_6$H$_4$	6	91

The effect of the alkali metal hydroxide on the reaction efficiency of the benzilic acid rearrangement in the solid state was different from that in solution. The effect on the reaction efficiency of the rearrangement of **140a** in the solid state increased in the following order: KOH>Ba(OH)$_2$>NaOH>CsOH (Table 31). On the other hand, the reaction efficiency of rearrangement of **140a** in boiling 50% aqueous EtOH increases in the order: KOH>NaOH>LiOH>Ba(OH)$_2$>CsOH. The rearrangement using Ba(OH)$_2$ proceeds faster in the solid state than in solution. However, LiOH is inert to the solid-state rearrangement, although it is effective in solution. The benzilic acid rearrangement has also been found to proceed via a radical intermediate as in solution. For example, a freshly prepared mixture of finely powdered **140e** and KOH showed a strong ESR signal (g=2.0049) and the signal declined as the reaction proceeded.

Table 31   Effect of alkali metal hydroxide on the benzilic acid rearrangement of **140a**

Alkali metal hydroxide	Reaction time / h	Yield / % of **141a**	
		Solid state	50% aq. EtOH
LiOH	6	0	70
NaOH	1	83	91
KOH	0.2	90	95
CsOH	5	89	43
Ca(OH)$_2$	6	0	trace
Ba(OH)$_2$	0.5	89	70
Al(OH)$_3$	6	0	trace

3.25. Meyer–Schuster rearrangement

Toluene-p-sulfonic acid (TsOH)-catalyzed Meyer–Schuster rearrangement of propargyl alcohols **101** also occurs in the solid state [18]. Keeping a mixture of **101** and TsOH at 50 °C for 2–3 h gives the aldehydes **142** in the yields shown in Table 32.

$$R^1R^2C\text{-}C\equiv CH \xrightarrow[\text{solid}]{\text{TsOH}} R^1R^2C=CHCHO$$
$$\underset{\text{OH}}{|}$$
101   **142**

Table 32   TsOH-catalyzed Meyer–Schuster rearrangement of **101** at 50 °C in the solid state

	R^1	R^2	Reaction time (h)	Yield (%)
a	Ph	Ph	2	58
b	Ph	o-Cl-C$_6$H$_4$	3	60
c	2,4-Me$_2$-C$_6$H$_3$	2,4-Me$_2$-C$_6$H$_3$	3	94

3.26. Beckmann rearrangement

When a solution of racemic 4-methyl-1-(hydroxyimino)cyclohexane **143** and (–)-1,6-bis(o-chlorophenyl)-1,6-diphenylhexa-2,4-diyne-1,6-diol **11** in ether-petroleum was kept at room temperature, a 1:1 inclusion compound of (–)-**11** and (+)-**143** of 79%ee was obtained as colorless needles. The same treatment of (±)-**145** and (–)-**11** gave a 1:2 inclusion compound of (–)-**11** and (+)-**145** of 59%ee. Heating of the 1:1 inclusion compound of (–)-**11** and (+)-**143** of 79%ee or the 1:2 inclusion compound of (–)-**11** and (+)-**145** of 59%ee with concentrated H$_2$SO$_4$ gave (–)-5-methyl-ε-caprolactam **144** of 80%ee or (+)-cis-3,5-dimethyl-ε-caprolactam **146** of 59%ee, respectively [45].

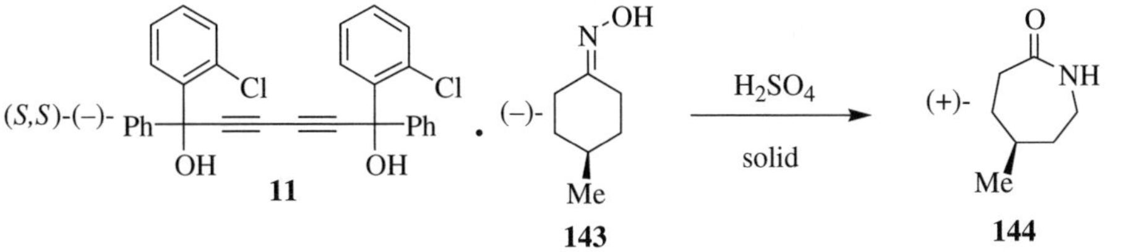

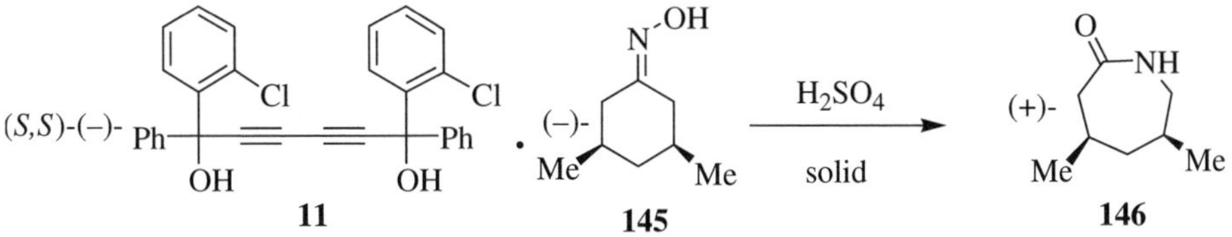

4. Organic Reaction in a Water Suspension Medium

Epoxidation of chalcones with NaOCl in a water suspension was found to proceed very efficiently [46]. For example, a mixture of **147a**, hexadecyltrimethylammonium bromide and commercially available 11% aqueous NaOCl was stirred at room temperature for 24 h. The reaction product was filtered and dried to give **148a** in quantitative yield. This procedure was applied to various kinds of chalcone derivatives, and **147b–i** were oxidized efficiently to give the corresponding epoxides **148b–i** in good yields (Table 33).

Bromination of crystalline powder of stilbene **20** with **25** in a water suspension medium proceeds selectively and efficiently [15]. For example, a suspension of both powdered **20** and **25** in a small amount of water was stirred at room temperature for 15 h. The reaction mixture was filtered and air dried to give *meso*-**21** in 88% yield. Bromination of chalcones **5** was also found to proceed very efficiently and selectively in a water suspension medium. For example, a suspension of powdered chalcone **5** and **25** in a small amount of water was stirred at room temperature for 1.5 h to give *erythro*-**23** in 90% yield.

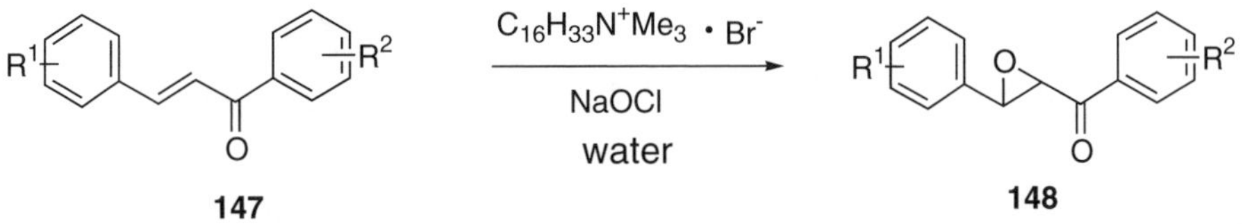

a: R^1=R^2=H

b: R^1=*m*-Me, R^2=H

c: R^1=*p*-Me, R^2=H

d: R^1=*p*-Cl, R^2=H

e: R^1=*p*-Br, R^2=H

f: R^1=*p*-MeO, R^2=H

g: R^1=H, R^2=*p*-Me

h: R^1=H, R^2=*p*-Cl

i: R^1=H, R^2=*p*-Br

j: R^1=H, R^2=*p*-MeO

k: R^1=R^2=*p*-Me

l: R^1=R^2=*p*-Cl

Table 33 Epoxidation reactions of chalcones in a water suspension medium

Chalcone	Product	Reaction time (day)	Yield (%)
11a	12a	1	100
11b	12b	2	80
11c	12c	2	85
11d	12d	4	85
11e	12e	2	90
11f	12f	1	78
11g	12g	2	36
11h	12h	0.4	90
11i	12i	0.3	99
11j	12j	1	93
11k	12k	5	43
11l	12l	5	30

Asymmetric bromination of 4,4′-dimethylchalcone **147k** in its chiral crystals was accomplished starting from optically inactive molecules. A powdered chiral crystal of **147k** prepared by recrystallization was exposed to bromine vapor for 2–3 h to give an optically active erythro-dibromide **149k** of 6%ee. The enantioselectivity of the asymmetric bromination of **147k** was found to improve when the reaction was carried out in a water suspension medium. For example, when the powdered chiral crystal of **147k**, which shows (–)-Cotton effect in the solid-state CD spectrum (Figure 10)[47], was stirred in a small amount of water containing of **25** for 3 h, optically active adduct (–)-**149k** of 13%ee was obtained in 73% yield.

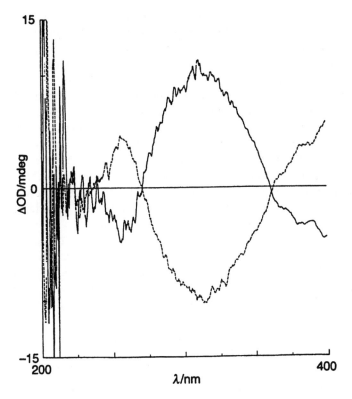

Figure 10 CD spectra of two enantiomeric chiral crystals of **147k** in Nujol mulls.

Michael addition reactions of amines, thiophenol and methyl acetate to chalcone in a water suspension medium also proceed very efficiently [48]. For example, a suspension of powdered chalcone **5** in a small amount of water containing n-BuNH$_2$ and a surfactant was stirred at room temperature for 4 h. The reaction product was filtered and air dried to give the Michael addition product **150** as a colorless powder in 98% yield. By the same procedure, Michael addition reactions of thiophenol to p-methoxychalcone **147f** in the presence of K_2CO_3 gave **151** in 92% yield. The Michael addition reaction of methyl acetoacetate to **5** also gave the addition product **152** in 98% yield.

The preparation of imines has been carried out by refluxing a mixture of amines and the carbonyl compounds in an organic solvent under azeotropic conditions in order to separate the water formed. It was found that the condensation reaction of benzaldehydes (**153**) and anilines (**154**) to benzylideneanilines (**156**) occurred efficiently in a water suspension medium without using any acid catalyst, and the products were isolated simply by filtration [49]. For example, a mixture of benzaldehyde and p-chloroaniline was stirred in a small amount of water at room temperature for 30 min. The crystalline powder formed was collected by filtration, washed with water and dried in a desiccator to give p-chlorobenzylideneaniline (**156c**) in 99% yield (Table 34).

Condensation reaction of aldehydes **153** and **N,N′**-disubstituted ethylenediamines **155** in a water suspension medium was also found to produce tetrahydroimidazole derivatives (**157**) very efficiently [49]. For example, a mixture of benzaldehyde and **N,N′**-diphenylethylenediamine was stirred in a small amount of water at room temperature for 30 min. The crystalline powder formed was collected by filtration, washed with

Table 34   Synthesis of benzylideneanilines (**156**) in a water suspension medium

	R^1	R^2	Reaction time (h)	Yield (%)	Mp (°C)
a	C_6H_5	C_6H_5	3	98	50–53
b	C_6H_5	p-MeC_6H_4	1	86	52
c	C_6H_5	p-ClC_6H_4	0.5	99	58–61
d	C_6H_5	p-BrC_6H_4	0.5	98	62–65
e	C_6H_5	p-$MeOC_6H_4$	3	86	66–69
f	p-MeC_6H_4	C_6H_5	1	88	88
g	p-ClC_6H_4	C_6H_5	2	97	58–61
h	p-BrC_6H_4	C_6H_5	1	86	71–74
i	p-PhC_6H_4	C_6H_5	0.5	92	155–158
j	p-HOC_6H_4	C_6H_5	0.5	94	50–53
k	p-ClC_6H_4	p-ClC_6H_4	0.5	87	110–113
l	m-HOC_6H_4	p-BrC_6H_4	0.5	94	111–114

Table 35   Synthesis of tetrahydroimidazoles (**157**) in a water suspension medium

	R^1	R^3	Reaction time (h)	Yield (%)	Mp (°C)
a	C_6H_5	C_6H_5	0.5	92	80–83
b	p-MeC_6H_4	C_6H_5	1.5	95	153–156
c	C_6H_5	p-MeC_6H_4	3	90	oil[a]
d	p-MeC_6H_4	p-MeC_6H_4	4	88	oil[a]
e	p-ClC_6H_4	p-MeC_6H_4	4	86	oil[a]

[a] The products were isolated by extraction with ether.

water and dried in a desiccator to give 1,2,3-triphenyltetrahydroimidazole (**157a**) in 92% yield (Table 35). This method is more convenient and efficient than the previously reported method in an organic solvent. It is very interesting that the formation of imines occurs efficiently using water as a solvent without the need for catalysis, or the azeotropic removal of water.

$$R^1{-}CHO \;+\; R^2{-}NH_2 \;\xrightarrow[\text{water suspension}]{-H_2O}\; \underset{H}{\overset{R^1}{\big\backslash}}C{=}N{-}R^2$$

$$\text{153} \qquad\quad \text{154} \qquad\qquad\qquad\qquad\qquad \text{156}$$

$$R^1{-}CHO \;+\; \left[\begin{array}{c}{-}NHR^3\\{-}NHR^3\end{array}\right. \;\xrightarrow[\text{water suspension}]{-H_2O}\; \text{157}$$

$$\text{153} \qquad\quad \text{155}$$

References

1. F. TODA, K. TANAKA and A. SEKIKAWA. *J. Chem. Soc., Chem. Commun.* (1987) 279.
2. F. TODA, K. TANAKA, H. MIYAMOTO, H. KOSHIMA, I. MIYAHARA and K. HIROTSU. *J. Chem. Soc., Perkin Trans.* **2** (1997) 1877.
3. F. TODA. *J. Synth. Org. Chem. Jpn.* **48** (1990) 494.
4. M. KAFTORY, K. TANAKA and F. TODA. *J. Org. Chem.* **50** (1985) 2154.
5. F. TODA, K. MORI, Y. MATSUURA and H. AKAI. *J. Chem. Soc., Chem. Commun.* (1990) 1591.
6. F. TODA and H. TAKUMI. *Enantiomer* **1** (1996) 29.
7. F. TODA, A. SATO, K. TANAKA and T. C. W. MAK. *Chem. Lett.* (1989) 873; G. KAUPP, J. SCHMEYERS, F. TODA, H. TAKUMI and H. KOSHIMA. *J. Phy. Org. Chem.* **9** (1996) 795.
8. K. TANAKA and F. TODA. *Chem. Rev.* **100** (2000) 1025.
9. F. TODA, M. YAGI and K. KIYOSHIGE. *J. Chem. Soc., Chem. Commun.* (1988) 958.
10. F. TODA, K. KIYOSHIGE and M. YAGI. *Angew. Chem. Int. Ed. Engl.* **28** (1989) 320.
11. M. EPPLE, S. EBBINGHAUS, A. RELLER, U. GLOISTEIN and K. CAMMENGA. *Thermochimica Acta* **269/270** (1995) 433. M. EPPLE and S. EBBINGHAUS. *J. Thermal Anal.* **52** (1998) 165.
12. F. TODA and K. MORI. *J. Chem. Soc., Chem. Commun.* (1989) 1245.
13. F. TODA and K. TANAKA. *Tetrahedron Lett.* **29** (1988) 551.
14. L. R. NASSIMBENI, M. L. NIVEN, K. TANAKA and F. TODA. *J. Cryst. Spect. Res.* **21** (1991) 451.
15. K. TANAKA, R. SHIRAISHI and F. TODA. *J. Chem. Soc., Perkin Trans.* **1** (1999) 3069.
16. H. HAMAZAKI, S. OHBA, F. TODA and H. TAKUMI. *Acta Cryst.* **C53** (1997) 620.
17. F. TODA, K. TANAKA and J. SATO. *Tetrahedron: Asymm.* **4** (1993) 1771.
18. F. TODA, H. TAKUMI and M. AKEHI. *J. Chem. Soc., Chem. Commun.* (1990) 1270; F. TODA, K. OKUDA. *J. Chem. Soc., Chem. Commun.* (1991) 1212.
19. F. TODA, K. TANAKA, T. TAMASHIMA and M. KATO. *Angew. Chem. Int. Ed.* **37** (1998) 2724.
20. K. TANAKA, N. TAKAMOTO, Y. TEZUKA, M. KATO and F. TODA. *Tetrahedron* (2001) in press.
21. G. KAUPP, J. SCHMEYERS, M. KATO, K. TANAKA, N. HARADA and F. TODA. *J. Phys. Org. Chem.* (2001) in press.
22. F. TODA, K. TANAKA and K. HAMAI. *J. Chem. Soc. Perkin Trans.* **1** (1990) 3207.

23. F. TODA, T. SUZUKI and S. HIGA. *J. Chem. Soc., Perkin Trans.* **1** (1998) 521.
24. K. TANAKA, T. SUGINO and F. TODA. *Green Chem.* **2** (2000) 303.
25. W. S. JOHNSON, C. E. DAVIS, R. H. HUNT and G. STORK. *J. Am. Chem. Soc.* **70** (1948) 3021.
26. F. TODA, H. TAKUMI and H. YAMAGUCHI. *Chem. Exp.* **4** (1989) 507.
27. H. TANAKA, S. KISHIGAMI and F. TODA. *J. Org. Chem.* **56** (1991) 4333.
28. F. TODA and H. AKAI. *J. Org. Chem.* **55** (1990) 3446.
29. F. TODA and N. IMAI. *J. Chem. Soc. Perkin Trans.* **1** (1994) 2673.
30. F. TODA and K. KANEMOTO. *Hetrocycles* **46** (1997) 185.
31. J. SCHMEYERS, F. TODA, J. BOY and G. KAUPP. *J. Chem. Soc., Perkin Trans.* **2** (1998) 989.
32. H. TANAKA, S. KISHIGAMI and F. TODA. *J. Org. Chem.* **55** (1990) 2982.
33. F. TODA, K. TANAKA and S. IWATA. *J. Org. Chem.* **54** (1989) 3007.
34. T. HIGASHIZIMA, N. SAKAI, K. NOZAKI and H. TAKAYA. *Tetrahedron Lett.* **35** (1994) 2023.
35. F. TODA and Y. TOKUMARU. *Chem. Lett.* (1990) 987.
36. K. TANAKA, H. AOKI, H. HOSOMI and S. OHBA. *Org. Lett.* **2** (2000) 2133.
37. E. C. HORNING. *Org. Syn., Coll.* **III** (1955) 281.
38. T. SUGINO and F. TANAKA. *Chem. Lett.* (2001) in press.
39. E. C. HORNING and M. G. HORNING. *J. Am. Chem. Soc.* **69** (1947) 968.
40. F. TODA, S. HYODA, K. OKADA and K. HIROTSU. *J. Chem. Soc., Chem. Commun.* (1995) 1531.
41. F. TODA and T. SHIGEMASA. *J. Chem. Soc., Perkin Trans.* **1** (1989) 209.
42. G. KAUPP, M. HAAK and F. TODA. *J. Phys. Chem.* **8** (1995) 545.
43. D. R. BOND, S. A. BOURNE, L. R. NASSIMBENI and F. TODA. *J. Cryst. Spectrosc. Res.* **19** (1989) 809.
44. F. TODA, K. TANAKA, Y. KAGAWA and Y. SAKAINO. *Chem. Lett.* (1990) 373.
45. F. TODA and H. AKAI. *J. Org. Chem.* **55** (1990) 4973.
46. F. TODA, H. TAKUMI, M. NAGAMI and K. TANAKA. *Hetrocycles* **47** (1998) 469.
47. F. TODA, K. TANAKA and M. KATO. *J. Chem. Soc., Perkin Trans.* **1** (1998) 1315.
48. F. TODA, H. TAKUMI, M. NAGAMI and K. TANAKA. *Hetrocycles* **47** (1998) 469.
49. K. TANAKA and R. SHIRAISHI. *Green Chem.* **2** (2000) 272.

Intermolecular Methyl Migration in the Solid State

Menahem Kaftory

Department of Chemistry, Technion – Israel Institute of Technology, Haifa 32000, Israel

1. Introduction

The conversion of ammonium cyanate to urea, discovered by Friedrich Wöhler in 1848, is widely considered to mark the beginning of organic chemistry as a separate field. However, it is not a well-known fact that this particular transformation occurs in the solid state. Chemical reactions in organic crystals are, therefore, certainly not new to the scientific community. However, for many years there was no systematic development of the subject and many solid-state reactions remained untouched and considered to be 'nature curiosities'. Organic chemical reactions were carried out in solutions and the development of this field was due to the progress in theories and experiments centered on molecular properties and reactivity. The difficulties associated with the understanding of solid-state reactions arise mostly from the fact that the reactivity of the system is a characteristic of the entire assembly of molecules and there were, at that time, no experimental and theoretical methods to explore the structure of such an assembly. The development of the technique of X-ray crystallography provided the means with which the link between the structure of the assembly of molecules and the solid-state reactivity could be established. The basic rules for solid-state photochemistry in the crystal were formulated [1] by using the term 'topochemical' which was coined by Hertel [2]. This rule states that *a reaction in the solid state occurs with a minimum amount of atomic or molecular movement*. Such a statement implies that there should exist an upper limit for such distances beyond which reaction can no longer occur. Nevertheless, many exceptions to this rule have been found and they were classified as non-topochemically controlled reactions. Major obstacles to the progress of the field are the lack of techniques that enable the study of the structures of short-lived intermediates and the shortage of examples of single-crystal-to-single-crystal transformation. Therefore most of the conclusions are based on the crystal structure of the *pre-* and *post*-reacted compound. There is no doubt that with a more comprehensive understand of packing and of topochemical effects, solid-state organic reactions could be planned and exploited in organic chemistry.

Besides the intrinsic and basic scientific significance, organic reactions in the solid state are attractive for a variety of reasons. They are useful as a means of synthesizing novel products that may be very difficult, if not impossible, to prepare by other means. For example, solid-state polymerization of acetylenic compounds [3–5] was a subject for research of many groups for many years. The recognition that chemical reactions might take place in the solid, the knowledge of the rate of solid-state reactions [6] are important for the pharmaceutical industry, because drugs should have a long shelf-life without undergoing chemical changes that might cause undesired clinical effects. Review articles [7,8] and a few books [9–11] summarizing the knowledge in the general area of chemical reactivity in molecular solids have been published recently.

F. Toda (ed.), Organic Solid-State Reactions, 47–67.
© 2002 *Kluwer Academic Publishers. Printed in Great Britain.*

Perhaps the most intriguing challenge in modern solid-state chemistry is the ability to design materials with preferred properties or desired chemical reactivity. There is a rapid advancement in the understanding and in establishing fundamental ideas towards the ability to control structure architecture. Gathering data and summarizing scattered information is a small but significant contribution towards this goal. In this chapter we summarize the important data relevant for the study of thermally activated intermolecular methyl transfer in the solid state.

2. Methyl Migration in the Solid State

Generally, the rate of monomolecular chemical reaction in the solid state is expected to be much slower than the rate of the same reaction when it is carried out in solution because of the rigidity imposed by the packing of molecules and the limited space available for molecular changes. The condition is expected to be reversed for bimolecular or multi-molecular reactions. There are special conditions for a reaction to be executed in solution or in the molten state, which is statistically limited or controlled. However, in the solid state in cases where the molecules are packed in a suitable arrangement that enable a reaction to take place, the reaction will proceed faster than in the other states of matter. Intermolecular methyl migration is an example of such reactions and the investigation of chemical systems undergoing that kind of reaction is summarized below.

The thermal rearrangement of aryl imidates to substituted amides (see **Scheme 1**), was shown to proceed by an intramolecular mechanism [12,13]. The thermal rearrangement is less favorable for alkyl imidates, which usually require very high reaction temperatures yet give low yields of substituted amide [14]. There are examples where the same reaction takes place in the solid state at much lower temperature with much higher yields.

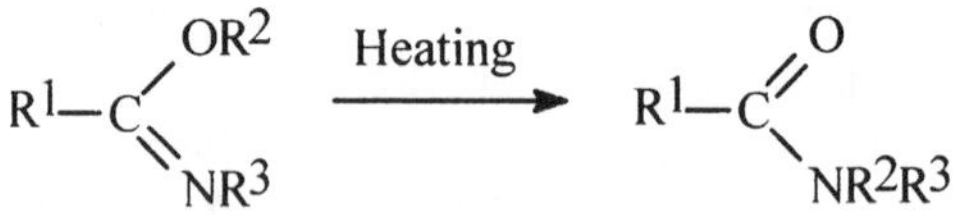

$$R^1 - C \overset{OR^2}{\underset{NR^3}{\diagup}} \xrightarrow{\text{Heating}} R^1 - C \overset{O}{\underset{NR^2R^3}{\diagup}}$$

Scheme 1

In the following sections we summarize the results of investigations of chemical systems undergoing intermolecular methyl transfer in the solid state. A summary of the reactions mentioned in the text is given in Table 1, while Table 2 includes the relevant geometrical data.

3. 4H-1,2,4-Triazoles

In the crystalline state the 4-methyl and 4-ethyl substituted 4H-1,2,4-triazoles (**1** in Table 1) were found to undergo thermal rearrangement to the corresponding 1-alkyl substituted triazoles [15] (**2** in Table 1). The crystal structures of the 4-methyl (**1a**) and the

Table 1 Summary of compounds undergoing methyl rearrangement

Original compound	Product	Reaction condition	Reaction temp. (°C)	Comments
(structure 1) R=Me, Et **1**	(structure 2) R=Me, Et **2**	Solid state	R=Me m.p. 160 R=Et m.p. 245	Topochemically controlled
(structure 3) **3**	(structure 4) **4**	Solid state	m.p. 214	Topochemically controlled
(structure 5) **5**	(structure 6) **6**	Solid state	Reaction temperature 95–145	Topochemically controlled
(structure 7) **7**	(structure 9) **9**	Solid state	Measured at 110	Crystal structure is unknown
(structure 10) **10**	(structure 11) **11**	Solid state	Measured at 132, 150	Crystal structure is unknown
(structure 14) **14**	(structure 15) **15**	Solid state	R=Me X=Ome Measured at 60–80.5 m.p. 122	Topochemically controlled
(structure 16) **16**	(structure 17) **17**	Solid state	Measured at 36–49 m.p. 54	Non-topochemically controlled

Table 1 Continued

Original compound	Product	Reaction condition	Reaction temp. (°C)	Comments
18	**19**	Solid state	Measured at 81, 88 m.p. 90	Topochemically controlled
20	A mixture of three products **21, 22, 23**	Solid state	Measured at 83–115	Crystal structure is unknown
29	**32**	Liquid state	Reaction temperature 173 m.p. 110	
30	**32**	Solid state	112	Topochemically controlled
31	**32**	Liquid state	Reaction temperature 1 m.p. 100–112	
33	**34**	Solid state	80	Topochemically controlled
35	**36**	Solid state	103	Topochemically controlled

Table 1 Continued

Original compound	Product	Reaction condition	Reaction temp. (°C)	Comments
37	Three products were isolated	Solid state	125	Non-topochemically controlled
41	Mixture (not isolated)	Liquid state	Reaction temperature 185 m.p. 130	
42	Mixture (not isolated)	Liquid state	Reaction temperature 255 m.p. 105	
43	Mixture (not isolated)	Liquid state	Reaction temperature 200 m.p. 170	
44	Few products were isolated 45,46,47	Solid state	205	Non-topochemically controlled

4-ethyl (**1b**) substituted triazoles are shown in Figures 1 and 2, respectively. The relevant geometrical parameters are given in Table 2.

Kinetic investigation of the thermolysis of both compounds, in the solid and in the liquid state was undertaken. It was found out that the activation energies for both compounds were higher in the crystals than in the melts. Also, the activation energy of the

Table 2 Relevant geometrical parameters

Comp.	X	Y	$Z_1 Z_2$	d1 (Å)	d2 (Å)	$\alpha(°)$	$\beta_1, \beta_2(°)$
1a	N	N	N C	1.464	3.242	167.1	77.6 120.3
1b	N	N	N C	1.481	4.610	161.0	152.3 94.4
3	O	N	C C	1.433	3.653	105.5	128.5 115.3
5	O	N	H H	1.466	3.198	173.2	107.6
14	O	N	N C	1.601	2.922	164.2	108.3 149.2
18	O	N	C C C	1.413	3.469	150.9	105.0 84.7 85.3
29	O	N	C C	1.444 1.446 1.446	3.553 3.587 3.356	115.9 124.9 118.6	98.7 148.2 / 117.9 129.0 / 100.9 146.1
30	O	N	C C	1.464 1.466 1.459	3.574 3.580 3.616	143.1 128.9 133.7	96.0 146.5 / 96.5 149.0 / 100.6 156.5
31	O	N	C C	1.453	3.383	154.8	96.8 144.8
33	O	N	C C	1.470	3.733	132.0	105.0 137.3
33	O	S	C	1.440	3.349	144.7	176.1
35	O	S	C	1.457 1.452	3.782 3.855	172.5 114.0	127.5 89.7
37	O	N	C C	1.460 1.447	3.686 3.817	116.7 116.5	86.4 145.7 / 90.3 150.0
41	O	N	C C	1.462	3.411	150.0	94.8 153.5
42	O	N	C C	1.443	3.738	111.0	115.4 131.7

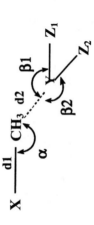

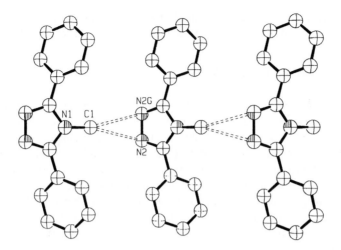

Figure 1   Crystal structure of **1a**.

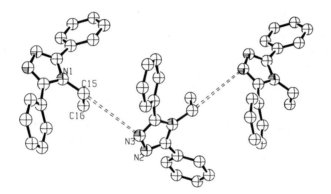

Figure 2   Crystal structure of **1b**.

ethyl derivative ($173 \, \text{kJ} \, \text{mol}^{-1}$) was higher than for the methyl derivative ($104 \, \text{kJ} \, \text{mol}^{-1}$). This agrees with the crystal structure since the interatomic distance between the reacting centers is shorter ($3.242 \, \text{Å}$) in the first than in the latter ($4.610 \, \text{Å}$). Thus, it seems that **1a** is closer to the ideal geometry for an $S_N 2$-type group transfer. The calculated activation entropy of **1a** ($\Delta S^{\#} = -115 \, \text{J} \, \text{mol}^{-1} \, \text{K}^{-1}$) is much lower than that of **1b** ($\Delta S^{\#} = -32 \, \text{J} \, \text{mol}^{-1} \, \text{K}^{-1}$). The activation energies for the same rearrangement in the melt found to be lower ($89 \, \text{kJ} \, \text{mol}^{-1}$ for **1a** and $129 \, \text{kJ} \, \text{mol}^{-1}$ for **1b**). The calculated activation entropies in the melt found to be smaller than those in the solid ($-124 \, \text{J} \, \text{mol}^{-1} \, \text{K}^{-1}$ for **1a**, $-58 \, \text{J} \, \text{mol}^{-1} \, \text{K}^{-1}$ for **1b**). Although a good agreement with first-order kinetics was found, it does not necessarily suggest that the reaction is unimolecular, but may rather be the result of a more complex reaction. The kinetic thermodynamic and structural data can be rationalized by the assumption that the rate-determining step of the reaction takes place between neighboring molecules in the crystal. It might be explained by a process where a dialkyltriazonium triazolate initially formed and later acts as a catalyst. The ion thus formed reacts with close neutral molecules to form the 1-alkyltriazole.

4. 4,8-Dimethoxy-1,5-naphthyridine

When **3** (see Table 1) is heated to 226 °C for 10 hr. it gives **4** in 62% yield [16]. Stereoscopic drawing of the crystal structure [17] is shown in Figure 3, and the relevant geometrical parameters are given in Table 2. The shortest distance between the methyl-carbon atom (C7) to the nitrogen atom (N1) of a second molecule is 3.653 Å. The methyl transfer could take place as shown in Figure 3. However, it should be noted that there is a second possible course (not shown) where the same carbon atom (C7) may migrate to a nitrogen atom of a molecule related to the one shown in Figure 3 by translation along the *c*-axis (perpendicular to the drawing). The distance between the two atoms is 3.986 Å. No further work has been done with regards to the study of the solid-state methyl rearrangement.

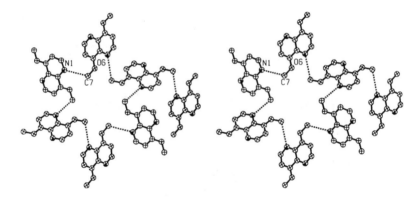

Figure 3   Stereoscopic view of the crystal structure of **3**.

5. (±)-2-Amino-2′-hydroxy-3′-(methoxycarbonyl)-1,1′-binaphthyl

When crystals of (±)-2-amino-2′-hydroxy-3′-(methoxycarbonyl)-1,1′-binaphthyl (**5** in Table 1) were heated [18] at 150 °C for 15 min, almost quantitative conversion occurred to the carboxylic acid (**6** in Table 1). The solid-state reaction was monitored by solid-state ^{13}C NMR spectroscopy. The reaction is exothermic with 4.44 kcal mol^{-1} as revealed by differential scanning calorimetry [18]. It can be assumed that the reaction occurs via the S_N2 pathway. An 'ideal' geometry for this mechanism was found between two molecules of the same handedness (see geometrical parameters in Table 2, and the crystal structure in Figure 4). The reaction is therefore enantiospecific.

Rigorous proof was provided by double isotopic labeling with CD_3 in one enantiomer, and ^{18}O in the other. The two enantiomers were then crystallized in a racemic mixture. Mass spectrometric analysis of the product confirmed the enantioselective process predicted by the data from the crystal structure.

O3
C22
N1

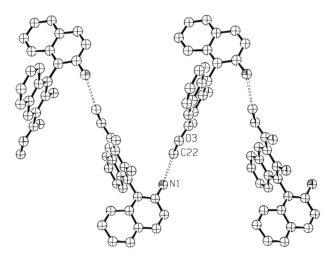

Figure 4 Crystal structure of **5**.

6. Benzisoxazole- and Naphthisoxazolequinones

Benzisoxazolequinone **7** and naphthisoxazolequinone **10** (see **Schemes 2** and **3**) undergo thermal rearrangement in solution and in the solid state [19]. In solution of **7** the major product obtained upon heating, is γ-cyanomethylidenebutenolide **8** and the corresponding N-methylisoxazolonequinone **9** was not observed. In the solid state on the other hand, the major product is **9**, and only small amount of **8** was detected (except for **7e**, where the major product was **8e**).

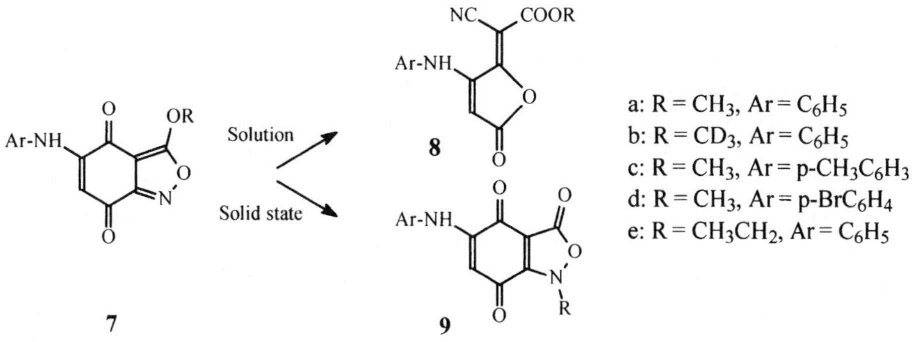

Ar-NH
O
OR
O
N
O
Solution
Solid state
NC
COOR
Ar-NH
O
O
Ar-NH
O
O
O
N
R
O
a: R = CH3, Ar = C6H5
b: R = CD3, Ar = C6H5
c: R = CH3, Ar = p-CH3C6H3
d: R = CH3, Ar = p-BrC6H4
e: R = CH3CH2, Ar = C6H5

7 **8** **9**

Scheme 2

The thermal rearrangement of **10** in a solution reveals three products; **11**, **12**, and **13**. The thermal rearrangement of **10** in the solids reveals only **11**. The result of this investigation clearly demonstrates the specificity of solid-state reaction. It also suggests that the solid-state reaction is topochemically controlled. Unfortunately the crystal structures of **7** and **10** are not known to confirm this assumption.

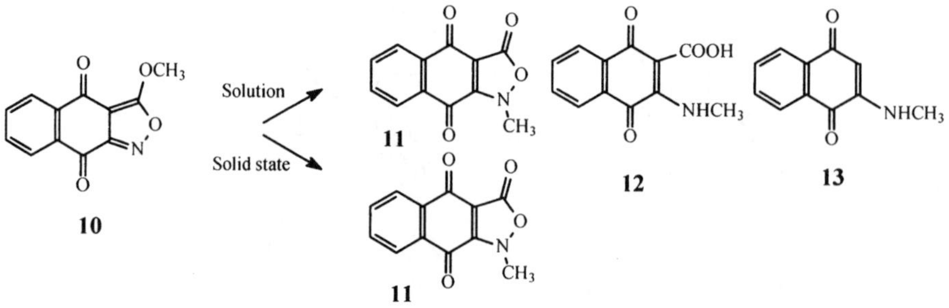

Scheme 3

7. 5-Methoxy-2-aryl-1,3,4-oxadiazoles

5-Methoxy-2-aryl-1,3,4-oxadiazoles (**14** in **Scheme 4**, see also Table 1) is a remarkable example of a family of compounds that undergo a Chapman-like rearrangement, which occurs much faster in the crystal than in the melt [20].

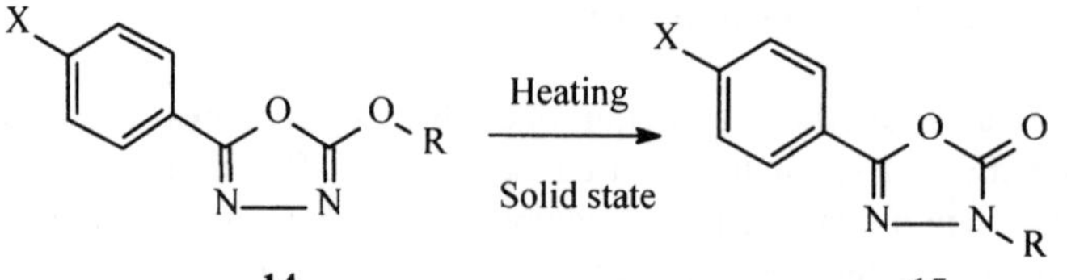

a: R= CH_3, X= H
b: R= CH_3, X= Cl
c: R= CH_3, X= OCH_3
d: R= CH_3, X= NO_2
e: R= CD_3, X= OCD_3

Scheme 4

The Chapman-like rearrangement is always difficult and requires temperatures higher than 200 °C. In contrast, the rearrangement takes place exceptionally easy for **14** when R=CH_3. The rearrangement takes place in the melt, at temperatures 120–140 °C and also takes place even more rapidly in the crystalline state. Compounds **14b–d** were found to be rearranged without apparent change in the crystalline state, within two years at room temperature. The crystal structure [21] of **14e** (see Figure 5 and geometrical data in Table 2) shows that the reaction is topochemically assisted. For example, the interatomic distance between the methyl-carbon atom and the closest nitrogen atom of a neighboring molecule is 2.922 Å (0.5 Å shorter than the sum of Van der Waals radii). The O–CD_3 bond length is 1.601 Å, significantly longer than a regular covalent bond (1.383, 1.407 Å). O–C$\cdots$N angle is 164.2°, very close to the ideal geometry for S_N2 reaction mechanism. It seems that the structure is an incipient stage of the rearrangement.

A theoretical study was undertaken [21], using model molecules. The transition structure for the methyl exchange was localized with an analytical gradient. The energies were calculated at the restricted Hartree–Fock level with 3-21G basis set. The computations suggest that the mechanism involved with the formation of the anion A^- and the cation B^+ followed by the transfer of a methyl group between these ions and the adjacent molecules in the same plane (see **Scheme 5**).

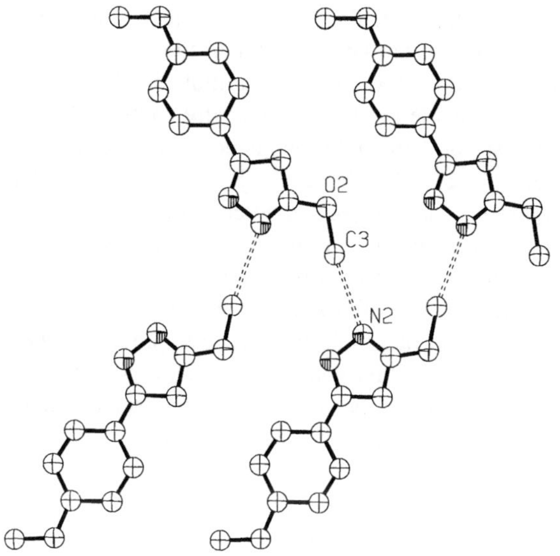

Figure 5 A layer in the crystal structure of **14e**.

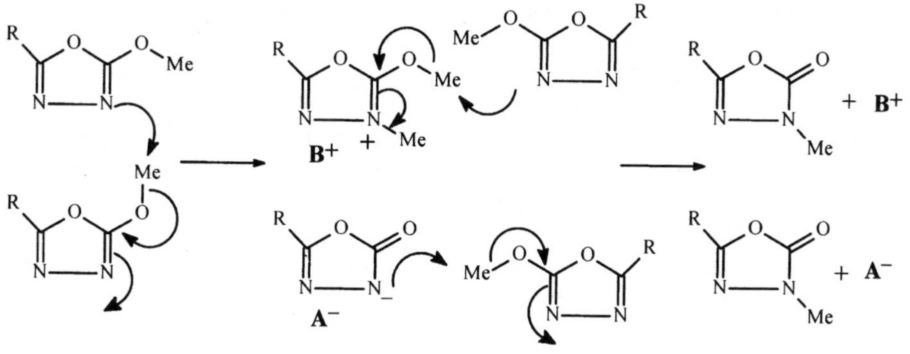

Scheme 5

A kinetic study using nuclear magnetic resonance (NMR) measurements [21] of this rearrangement supports this mechanism. The experimental activation energy for **14c** is $74.3\,kJ\,mol^{-1}$ for the reaction in the solid state, $122.3\,kJ\,mol^{-1}$ for the reaction in the melt and $85.4\,kJ\,mol^{-1}$ (in the liquid state with a salt of tetrafluoroborate) and $97.7\,kJ\,mol^{-1}$ (in the liquid state with a salt of a potassium) for the reaction of **14a**.

8. Methyl 2-(Methylthio)benzenesulfonate

The rearrangement of methyl 2-(methylthio)benzenesulfonate (see **16** in Table 1) to the zwitterionic 2-(dimethylsulfonium)benzenesulfonate (**17** in Table 1) is known to proceed in solution by intermolecular methyl transfers. The same rearrangement has been observed to occur in crystalline **16**, although the crystal structure [22] shows that the molecular packing is not conducive to intermolecular methyl transfer. The crystal structure does not look suitable for a topochemically controlled intermolecular S_N2 type mechanism where the sulfonyl methyl group is being transferred to the divalent S of the neighboring molecule; intermolecular $H_3C\cdots S$ distances are at least 5.25 Å, and the linear $O-C\cdots S$ requirement is not even remotely satisfied for any neighboring pair of molecules. The distinction between the inter- and intramolecular mechanisms for the solid-state methyl transfer reaction was made by analysis of the reaction product obtained from crystalline mixture of protio- and deuteriomethylated **16**. It was concluded that the methyl transfer is an intermolecular process. Since the crystal packing is quite unsuitable for intermolecular methyl transfer, one can infer that the reaction in the solid-state proceeds not topochemically but rather at defects such as microcavities, surfaces, and other irregularities in the ordered crystal arrangement.

9. p-Dimethylaminobenzenesulfonate

The conversion of *p*-dimethylaminobenzenesulfonate (**18** in Table 1) to *p*-trimethylamo-niumbenzenesulfonate zwitterions (**19** in Table 1), was first discovered by Kuhn and Ruelius [23]. Later, it was found [24] that the rate of the rearrangement increases with temperature below the melting temperature. In the melt, there is a sharp decrease of the reaction rate. Based on reaction rate measurements it appears that the solid-state rearrangement is at least 25 times faster than that in the melt. The distinction between intra- and intermolecular methyl rearrangement was established by a 'double label scrambling' experiment. The product of a reacted roughly 50:50 mixture of undeuter-ated and deuterated compound analyzed by desorption mass spectrometry. An extensive methyl scrambling was found in the product, indicating that the reaction is inter- rather than intramolecular.

The crystal structure at 193 K [25,26] is shown in Figure 6, relevant geometrical parameters are given in Table 2. From these data it can be seen that this rearrangement is topochemically assisted. The distance between methyl carbon atom to the nitrogen atom of a neighboring molecule (C9, and N1 in Figure 6) is 3.469 Å, and O–C⋯N angle is 150.9°, C9 approaching angle (C9⋯N1–C) is 105.0°. Moreover, the nitrogen atom deviates by 0.044 Å from planarity (of an sp^2 hybridization) towards the approaching methyl, thus developing its lone pair electron towards sp^3 hybridization. From crystal structure determination at three different temperatures [26] it was shown that the deviation of the nitrogen atom increases with the increase of temperature, to 0.07 Å at room temperature.

The mechanism of the rearrangement in the solid state was theoretically studied [27] by extended Huckel theory. Three different mechanisms have been checked: a one-step mechanism, a two-step mechanism, and a chain-reaction mechanism. Based on the calculation it was concluded that the two-step mechanism involving an intermediate whose structure is probably a molecular ion-pair. It was farther concluded that the enhancement of the reaction rate in the crystal is due to an entropic factor arising from

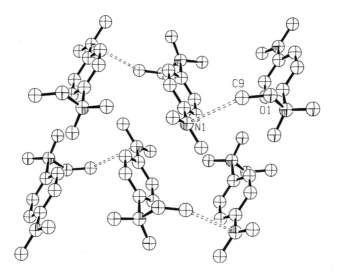

Figure 6   Crystal structure of **18**.

the favorable orientation of the reacting molecules, which easily overrides the extra activation energy due to crystal field effects. Raman phonon spectroscopy, which contains
both structural and dynamical information, was used for the study of the rearrangement
in the solid-state [28]. The result is consistent with the proposed mechanism that pre-
dicts a long-range cooperative rearrangement [26], rather than a local rearrangement. It
was also concluded that mode softening may play an important role in the thermal reaction. A more recent study [29] of the solid-state rearrangement by Raman spectroscopy
led to the conclusion that the presence of a soft mode is not prerequisite for assisting a
thermally enhanced solid-state reaction. In another study by diffuse reflectance IR, a
comparison has been made between the rearrangement in the bulk and that at the surface
[30]. It was proved that the reaction in the bulk is faster than that at the surface. It was
shown that the surface reaction reveals a mixture of the zwitterion and unidentified
product.

10.   Tetraglycine Methyl Ester

The chemical kinetics of thermally induced methyl transfer in tetraglycine methyl ester
[31] (**20**) was thoroughly investigated [32] by following the change in the concentration
of reactant and various intermediates and products at different temperatures by HPLC.
The phase changes during the reaction were followed by X-ray diffraction. Three main
products (**21–23**) and an intermediate (**24**) have been identified and it has been shown
that this reaction is accompanied by crystallization of a product phase [31]. This reaction is believed to involve intermolecular methyl transfer rather than intramolecular
rearrangement.

$$H_2N-(CH_2-\underset{\underset{20}{\parallel}}{\overset{O}{C}}-NH)_3 CH_2-\overset{O}{\overset{\parallel}{C}}-O-CH_3$$

$$H_3C-\underset{\underset{H}{\overset{+}{N}}}{\overset{H}{\overset{|}{N}}}-(CH_2-\underset{\underset{21}{\parallel}}{\overset{O}{C}}-NH)_3 CH_2-\overset{O}{\overset{\parallel}{C}}-O^- \qquad H_3\overset{+}{N}-(CH_2-\underset{\underset{22}{\parallel}}{\overset{O}{C}}-NH)_3 CH_2-\overset{O}{\overset{\parallel}{C}}-O^-$$

$$H_3C-\underset{\underset{CH_3}{\overset{+}{N}}}{\overset{H}{\overset{|}{N}}}-(CH_2-\underset{\underset{23}{\parallel}}{\overset{O}{C}}-NH)_3 CH_2-\overset{O}{\overset{\parallel}{C}}-O^- \qquad H_3C-\underset{\underset{H}{\overset{+}{N}}}{\overset{H}{\overset{|}{N}}}-(CH_2-\underset{\underset{24}{\parallel}}{\overset{O}{C}}-NH)_3 CH_2-\overset{O}{\overset{\parallel}{C}}-O-CH_3$$

DSC measurements were used to monitor the possible appearance of a liquid phase during the rearrangement, however, no apparent eutectic melting occurred. Moreover, there were no indications of a glass transition associated with an amorphous phase [31]. It was concluded that the reaction consists of two major parts, one involving a single-phase reaction and the second involving heterogeneous reaction. In the first part, a solid solution of products is formed in the reactant. This part was treated by a first-order kinetic model assuming a linear-concentration dependence of the rate constant. In the second part of the reaction the process controlled by phase transformation. The activation energies obtained for the two stages in the range of temperatures 83–115 °C are 100–130 kJ mol^{-1}. Unfortunately the crystal structure of **20** is not known and therefore there is no data to assist in concluding whether the reaction is topochemically or non-topochemically controlled.

11. Methyl Esters of Cyanuric Acid

Over 120 years ago, Hoffman and Olshausen [33] and later Klason [34] gathered evidence to prove that methyl esters of cyanuric acid and their thio derivatives undergo alkyl migration. From the end of the 1960s to the mid-1980s, the methyl rearrangement in methyl esters of cyanurates and thiocyanurates (**25–28** where X=O or S) was investigated [35,36]. Theoretically, there are 24 different isomers of methyl cyanurates and thiocyanurates. Most of them have been prepared [35,37–42]. Some of them were shown to undergo methyl rearrangement not only in solution or in the liquid state but also in the solid state.

25   **26**   **27**   **28**   X = O, S

The conversion of oxo-cyanurate (**29** in Table 1) in the melt was thoroughly investigated and a mechanism was proposed [35] to explain the formation of **32** (see Table 1) via the intermediates **30** and **31** (see Table 1) by O→N methyl migration. By appropriate labeling it was shown beyond doubt that the rearrangement is intermolecular. It was also proposed and proved that the rearrangement occurs as the consequence of the attack of Me^+ cations on the ring nitrogen of the neutral molecules. The presence of such cations in the melt was established by ionic exchange with the H^+ of the molten acids. The same mechanism was also observed for the reaction in solution. A homolytic process that will form methyl radical was ruled out because of the absence of side-products or decomposition products. Since the mean free path of Me^+ is very small, and that of the anions formed is even smaller, the exchange of methyl groups takes place at very short distances. The very fast rapid conversion of the intermediates can be regarded as a chain propagation of steps between several neighboring molecules once the first Me^+ was present. The slowing down of the overall reaction after 67% conversion, is an indication that the diffusion of the cation is limited to very short distances while the average intermolecular distances increase with the decrease of concentration.

Kaftroy and co-workers [39,42] have studied the relation between the crystal structures of methoxy traizines **29–31, 33, 35, 41–44**, their thermal behavior with respect to methyl rearrangement. The products obtained, the state of matter and the temperature of the rearrangement, and the relevant geometrical parameters are given in Tables 1 and 2. One can distinguish between three different types of rearrangements: a) liquid state b) topochemically controlled solid state, and c) non-topochemically controlled solid state. The first two types are easily detected while the last one is not. The distinction between the liquid- and solid-state rearrangements is based on differential scanning calorimetry (DSC). In the first, an endothermic peak is observed prior to the exothermic rearrangement peak (for example, see Figure 7) in the second case, an exothermic peak is the first to observed (see for example Figure 8). The distinction between the two types of solid-state rearrangement (b) and (c) is more complex. However, an example is shown in Figure 8. In the example shown on the right, the first exothermic peak is an indication for more than a single step rearrangement and it is followed by a sharp endothermic peak of a melting of a single product. In the example

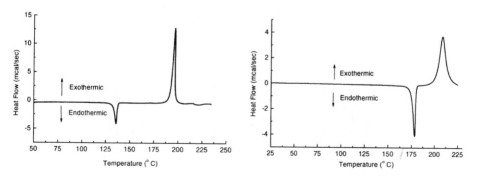

Figure 7   DSC thermographs of **41**, and **43**, at heating rate of $3\,°C\,min^{-1}$.

shown on the left, there is no indication for melt. It was found that in type (c) rearrangement several products are obtained and therefore there is no sharp melting endotherm (for discussion see below).

When a solid-state reaction is detected, it is useful to check the crystal structure of the reacting compound to find the relation between the structure and the reactivity. In most cases the relation is easily observed. Would it also be able to do the opposite, namely to be able to predict that a reaction will take place in the solid, based on the crystal structure? The experience gained during the study of the methyl rearrangement of methoxy triazines provides a clear answer, no! The crystal structure of **29** seems to be 'ideal' for topochemically controlled rearrangement (see Figure 9). The methyl groups are in the

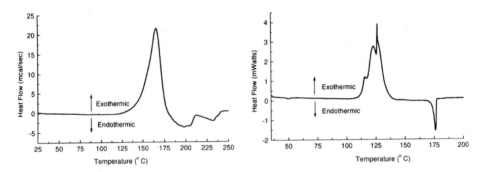

Figure 8   DSC thermographs of **37**, and **30**, at heating rate of $3\,^{\circ}\mathrm{C\,min^{-1}}$.

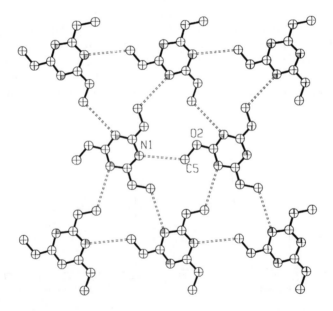

Figure 9   A layer of **30**.

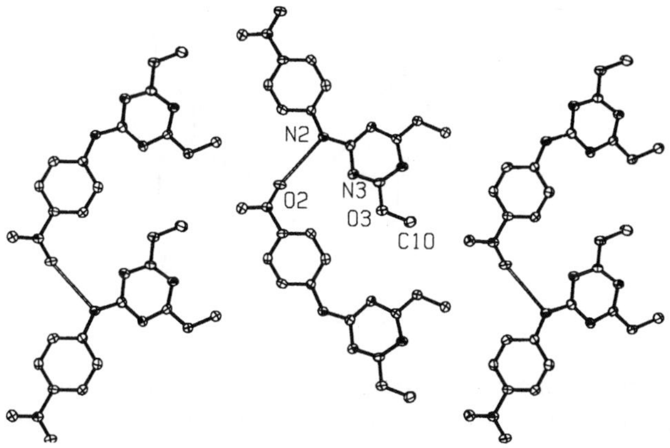

Figure 10   A layer of **44** (open bonds showing hydrogen bonds).

same plane of the neighboring nitrogen lone-pair electrons. The distances between the methyl carbon and these nitrogen atoms ranged between 3.536 and 3.587 Å, the approaching angle (O–C···N) range between 115.9 to 124.9°. Nevertheless, the rearrangement takes place in the melt. A competing phase transition at lower temperature (67 °C) occurs and a different structure is formed [39] which is no more 'ideal'. On the other hand, there are examples where the crystal does not have the structural properties expected for a potentially reactive solid. For example, although compound **44** (see Figure 10) adopts a layer structure, the methyl-carbon atom is closer to a nitro-oxygen atom (3.031 Å) than to a nitrogen atom of a neighboring molecule (4.175 Å), and the O–C···N angle is 85.0°, far from the ideal 180° expected for an S_N2 mechanism. The fact that this reaction is non-topochemically controlled is also evident from the relative high temperature of its execution (205 °C), and also because it produces a mixture of different compounds (see **Scheme 6**). A similar example is the solid-state rearrangement of **37** (see Table 1, and **Scheme 7** and Figure 8 left).

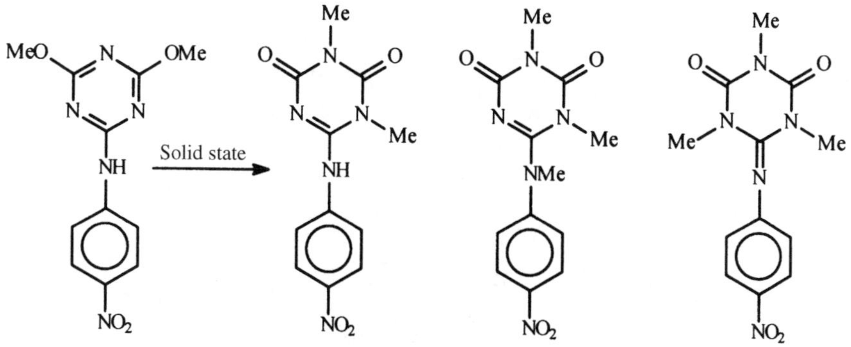

Scheme 6

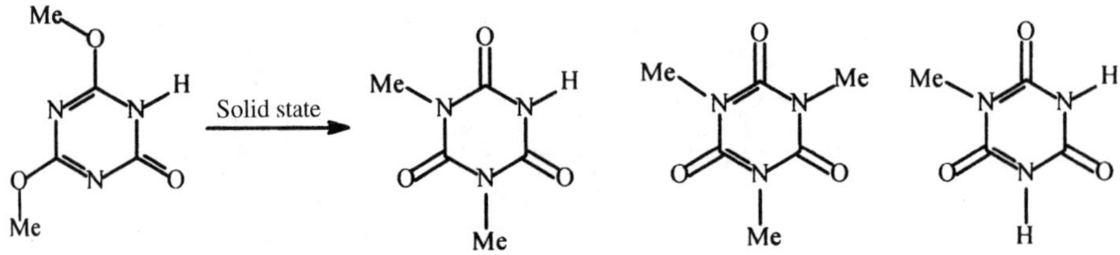

Scheme 7

32   33   34   35

In both cases the rearrangement products cannot be predicted, based on the crystal structure [42], so that it is unlikely to conclude that the rearrangement is topochemically controlled.

The most interesting type of rearrangement is the topochemically controlled reaction. **30** undergo two courses of rearrangements, a one-methyl rearrangement to yield **31**, and a two-methyl rearrangement to yield **32**. These results are based on both crystal structure and on kinetic studies [39]. The structure consists of parallel layers (see Figure 11 for a typical layer, and Figure 8 right, for the DSC thermograph) with two molecules in the asymmetric unit and they are inclined by approximately 7° to each other. There are short distances between the methoxy methyl-carbon atoms and the neighboring nitrogen atoms (3.574–3.615 Å) and O–C···N angles of the range 128.9 to 143.1°. However, it should be mentioned that a phase transition occurs before the rearrangement takes place, where the molecules go into a common plane [43].

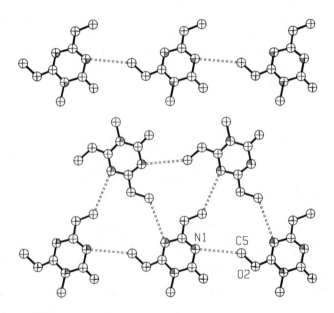

Figure 11   A layer of **30**.

Two similar compounds where sulfur atoms replace some of the oxygen atoms (**33** and **35**) [43] show somewhat different rearrangement in the solid state affording **34**, and **36**, respectively. In both, the methyl group is migrating to a sulfur atom of the neighboring molecule (O→S). Moreover, in **33**, although the distance between the methyl-carbon atom and a nitrogen atom of neighboring molecule is 3.733 Å, which in the other examples discussed above led to O→N migration the methyl is being transferred to the sulfur atom which is only 3.349 Å apart.

12. Summary

The examples of intermolecular methyl transfer discussed above may be divided into three different classes: solid-state topochemically controlled reaction, solid-state non-topochemically controlled reaction, and liquid-state reaction. While the first is specific and assisted by the crystal structure and therefore takes place at lower temperatures, the second is a non-specific reaction, it takes place at higher temperatures and it resembles the liquid-state reaction.

The mechanism of the solid-state methyl transfer was studied by a few groups. The main difference between the proposals is the initial step of the reaction. Gavezzotti and Simonetta [27] suggested that the first step in the rearrangement of **18** is the reaction between two molecules resulting in a pair of cation and anion (by transfer of a methyl group). This step is followed by the reaction between the two to form two zwitterionic molecules. A somewhat different mechanism was proposed for the rearrangement of **14** [21]. The first step is the S_N2 reaction between a nitrogen lone-pair of one molecule and a methyl carbon atom of a second molecule. The resulted molecular cation (the molecule that gained a methyl cation) and molecular anion (the molecule that lost a methyl cation) are then reacting with neutral molecules to form rearranged products and a second pair of molecular cation and molecular anion. The two proposed mechanisms are based on theoretical calculations. The latter was supported by kinetic measurements. Yet another mechanism was proposed [35] for the rearrangement of **29** based on the liquid-state kinetic measurements. The first step in the rearrangement is the dissociation of a neutral molecule to an anion and a methyl cation. The second step is the attack of the methyl cation on a neutral molecule to produce a molecular cation. The successive steps are not relevant for comparison because this special reaction involved more than a single methyl transfer. In all the examples discussed above, no experimental evidence for intermediates have been obtained. However, during the kinetic study of the rearrangement of **20** [32] experimental evidence for the formation of the intermediate molecular cation consisting of a neutral molecule bonded to a methyl cation was obtained. Unfortunately, the crystal structure of the parent compound is not available and therefore there is no proof that the reaction is topochemically controlled.

Activation energies for some of the reactions have been calculated from the data obtained from kinetic measurements. It is interesting to find that albeit there are differences between the chemical systems, the activation energies are similar. It is also interesting to compare the activation energies of solid-state and liquid-state reactions. The activation energy for the rearrangement [15] of **1a** is 104 kJ mol^{-1} in the solid state while it is $89 \, \text{kJ} \, \text{mol}^{-1}$ in the liquid state. The activation energies for the rearrangement [21] of **14c** in the solid state is lower ($74.3 \, \text{kJ} \, \text{mol}^{-1}$) then that in the liquid state ($122.3 \, \text{kJ} \, \text{mol}^{-1}$), and even smaller than that of a catalyzed reaction measured for **14a** as $85.4 \, \text{kJ} \, \text{mol}^{-1}$ (in the liquid state with a salt of tetrafluoroborate) and $97.7 \, \text{kJ} \, \text{mol}^{-1}$ (in the liquid state with a salt of a potassium). The activation energy for the solid-state rearrangement [32] of **20** was estimated to be $100\text{–}130 \, \text{kJ} \, \text{mol}^{-1}$. The activation energy of the solid-state rearrangement [39] of **30** was estimated to be $106.3 \, \text{kJ} \, \text{mol}^{-1}$ while that of the rearrangement [35] of **29** in the liquid state was calculated to be $118.3 \, \text{kJ} \, \text{mol}^{-1}$.

The intermolecular methyl transfer shown in this chapter indicates the advantages of solid-state reactions whenever the reaction is topochemically controlled because it is not only specific, leading to a single product, but also because the reaction is enhanced due to an entropic factor arising from the favorable orientation of the reacting molecules, which easily overrides the extra activation energy due to crystal field effects.

References

1. M. D. COHEN and G. M. J. SCHMIDT. *J. Chem. Soc.* (1964) 1996.
2. Z. HERTEL. *Elektrochem.* **37** (1931) 536.
3. H. BASSLER. *Adv. Polym. Sci.* **63** (1984) 1; and references cited therein.
4. M. HASEGAWA. In *"Organic Solid State Chemistry"*, edited by G. R. DESIRAJU, (Elsevier, Netherland, 1987) p. 153; and references cited therein.
5. Y. KAI and Y. YAMAMOTO. In *"Reactivity in Molecular Crystals"*, edited by Y. OHASHI, (VCH Publishers, Tokyo, Japan, 1993) p. 277.
6. J. HALEBLIAN and W. MCCRONE. *J. Pharm. Sci.* **58** (1969) 911.
7. N. B. SINGH, R. J. SINGH and N. P. SINGH. *Tetrahedron* **50(22)** (1994) 6441.
8. G. R. DESIRAJU. *Solid State Ionics* **101 (Pt. 2)** (1997) 839.
9. E. BOLDYREVA and V. BOLDYREV. *"Reactivity of Molecular Solids"*, (John Wiley & Sons Ltd., Chichester, UK, 1999).
10. G. R. DESIRAJU. *"Organic Solid State Chemistry"* (Elsevier, Amsterdam, 1987).
11. Y. OHASHI. *"Reactivity in Molecular Crystals"* (VCH, Tokyo, Japan, 1993).
12. K. B. WIBERG and B. I. ROWLAND. *J. Am. Chem. Soc.* **77** (1955) 2205.
13. O. H. WHEELER, F. ROMAN and O. ROSADO. *J. Org. Chem.* **34** (1969) 966.
14. C. G. MCCARTY and L. A. GARNER. *"The Chemistry of Amidines and Imidates"* edited by S. PATAI. (Wiley, London, 1975) p. 189.
15. P. H. J. CARLSEN, K. B. JOERGENSEN, O. R. GAUTUN, S. JAGNER and M. HAAKANSSON. *Acta Chem. Scand.* **49(9)** (1995) 676.
16. S. B. BROWN and M. J. S. DEWAR. *J. Org. Chem.* **43(7)** (1978) 1331.
17. R. L. HARLOW and S. H. SIMONSEN. *Acta Cryst.* **B33** (1977) 2662.
18. M. SMARČINA, Š. VYSKOČIL, V. HANUŠ, M. POLÁŠEK, V. LANGER, B. G. M. CHEW, D. B. ZAX, K. H. VERRIER, T. A. CLAXTON and P. KOČOVSKY. *J. Am. Chem. Soc.* **118** (1996) 487.
19. M. V. MARTINEZ-DIAZ, S. RODRIGUEZ-MORGADE, W. SCHAEFER and T. TORRES. *Tetrahedron* **49(11)** (1993) 2261.
20. a. M. DESSOLIN, M. GOLFIER, A. GONTHIER-VASSAL and H. SZWARC. *Nouv. J. Chim.* **10(12)** (1986) 753.
    b. M. DESSOLIN and M. GOLFIER. *J. Chem. Soc., Chem. Commun.* **(1)** (1986) 38.
21. M. DESSOLIN, O. EISENSTEIN, M. GOLFIER, T. PRANGE and P. SAUTET. *J. Chem. Soc. Chem. Commun.* (1992) 132.
22. P. VENUGOPALAN, K. VENKATESAN, J. KLAUSEN, E. NOVOTNY-BREGGER, C. LEUMANN, A. ESCHENMOSER and J. D. DUNITZ. *Helv. Chim. Acta* **74(3)** (1991) 662.
23. R. KUHN and H. W. RUELIUS. *Chem. Ber.* **83** (1950) 420.
24. C. N. SUKENIK, J. A. P. BONAPACE, N. S. MANDEL, P. LAU, G. WOOD and R. G. BERGMAN. *J. Am. Chem. Soc.* **99(3)** (1977) 851.

25.   C. N. SUKENIK, J. A. P. BONAPACE, N. S. MANDEL, R. G. BERGMAN, P. LAU and G. WOOD. *J. Am. Chem. Soc.* **97(18)** (1975) 5290.

26.   a. J. A. R. P. SARMA and J. D. DUNITZ. *Acta Crystallogr.* **B46** (1990) 780.
      b. J. A. R. P. SARMA and J. D. DUNITZ. *Acta Crystallogr.* **B46** (1990) 784.

27.   A. GAVEZZOTTI and M. SIMONETTA. *Nouv. J. Chim.* **2(1)** (1978) 69.

28.   J. EVEN, M. BERTAULT, A. GIRARD, Y. DELUGEARD and Y. MARQUETON. *Chem. Phys. Lett.* **267(5,6)** (1997) 585.

29.   K. DWARAKANATH and P. N. PRASAD. *J. Am. Chem. Soc.* **102(12)** (1980) 4254.

30.   F. M. MENGER, H. B. KAISERMAN and L. J. SCOTCHIE. *Tetrahedron Lett.* **25** (1984) 2311.

31.   a. L. A. SLUYTERMAN and H. J. VEENENDAAL. *Recuil* (1952) 137.
      b. L. A. SLUYTERMAN and H. J. VEENENDAAL. *Recuil* (1952) 277.

32.   E. YU. SHALAEV, S. R. BYRN and G. ZOGRAFI. *Int. J. Chem. Kinet.* **29(5)** (1997) 339.

33.   A. W. HOFFMANN and O. OLSHAUSEN. *Ber.* **3** (1870) 269.

34.   P. KLASON. *J. Prakt. Chem.* **33** (1885) 116.

35.   L. PAOLINI, M. L. TOSATO and M. CIGNITTI. *Heterocyclic Chem.* **5** (1968) 533.

36.   a. M. L. TOSATO. *J. Chem. Soc. Perkin Trans. II.* (1979) 1371.
      b. M. L. TOSATO. *J. Chem. Soc. Perkin Trans. II.* (1982) 1321.
      c. M. L. TOSATO. *J. Chem. Soc. Perkin Trans. II.* (1984) 1593.

37.   M. KAFTORY and E. HANDELSMAN-BENORY. *Molec. Cryst. Liq. Cryst.* **240** (1994) 241.

38.   E. HANDELSMAN-BENORY, M. BOTOSHANSKY and M. KAFTORY. *Acta Cryst.* **C51** (1995) 2421.

39.   E. HANDELSMAN-BENORY, M. BOTOSHANSKY, M. GREENBERG, V. SHTEIMAN and M. KAFTORY. *Tetrahedron.* **56** (2000) 6887.

40.   H. TAYCHER, V. SHTEIMAN, M. BOTOSHANSKY and M. KAFTORY. *Acta Crystallogr.* **C56** (2000) 832.

41.   M. GREENBERG, V. SHTEIMAN and M. KAFTORY. *Acta Crystallogr.* **C56** (2000) 465.

42.   H. TAYCHER, M. BOTOSHANSKY, V. SHTEIMAN and M. KAFTORY. *Supramolecular Chem.* **13** (2001) 181.

43.   M. GREENBERG, V. SHTEIMAN and M. KAFTORY. *Acta Crystallogr.* **B57** (2001) 428.

Solid-State Ionic Reactions

Keiji Kobayashi

Department of Chemistry, Graduate School of Arts and Sciences, The University of Tokyo, Komaba, Meguro-ku, Tokyo, Japan

1. Introduction

Organic ionic reactions usually require the substrate to be dissolved in a liquid solvent. In view of ecological and economical considerations, the development of waste-free reactions, that is, solvent-free reactions, is more desirable. Although a variety of organic solid-state reactions have been investigated up to now, these are mostly photoreactions that involve radical intermediates or proceed via cyclo-additions [1]. Ionic reactions of organic compounds induced in the solid state have rarely been explored. Strictly speaking, the reactions that have been attributed to the ionic mechanism in solution chemistry have rarely been explored in the solid state. There has been almost no evidence so far provided for the solid-state ionic reactions. It is rather difficult to draw a definitive conclusion for the solid-state reactions to proceed via an ionic mechanism. Therefore, solid-state ionic reactions quoted in this chapter are rationalized by analogy with solution chemistry.

The main reason for the scarcity of solid-state ionic reactions is that these should be induced thermally, and molecular crystals melt before the onset of the reaction. The reaction occurs with retention of the solid state when the melting points of the starting compounds and products are sufficiently higher than the reaction temperature. Thus, salts are advantageous for the solid reactions. We recall that the historical reaction discovered by Wöher is a solid-state thermal reaction: ammonium cyanate, a salt, is transformed to urea, a molecular material of high melting point.

$$NH_4^+ \ OCN^- \xrightarrow[\text{solid}]{} H_2N-CO-NH_2$$

As seen in this historical example, the reaction in the solid state is usually associated with a one-component reaction starting from a single solid. This type of reaction is generally a rearrangement or isomerization, and mostly proceeds as a crystal-to-crystal transformation. In order to realize a two-component solid-state reaction in the crystalline phase, such as a substitution or addition reaction, it is required to preorganize the components in the crystalline phase. Crystal engineering of two-component solids is important to develop solid-state bimolecular reactions. Clathrate crystals, charge-transfer complexes, hydrogen-bonded molecular complexes, and solid solution may be candidates. However, thermally induced ionic reactions have rarely been demonstrated for the two-component solids in contrast to solid-state photoreactions. Preorganization with the intention of a particular reaction is still difficult at the present stage of crystal engineering. Therefore, the solid-state ionic reactions between two reactants or between a substrate and a reagent are usually carried out by grinding the mixture of the two solid components. The reaction involves reorganization of the crystalline lattice from the

F. Toda (ed.), Organic Solid-State Reactions, 69–107.

starting to the product phase. In some cases, the product phase turns to an amorphous state.

Another approach to the solid-state ionic reaction is the reaction induced by gas-solid contact [2]. The Kolbe–Schmitt reaction is a well-known example of a gas-solid reaction.

The crystal surface is exposed to gaseous reagent. Otherwise, it would be suggested that the reaction takes place by the diffusion of a vapor component through cavities between microcrystalline particles and through interior defects of the other component.

In this chapter, the three types of the above-mentioned reactions, that is, (1) single-solid reaction, (2) solid-solid reaction induced by grinding or contacts, and (3) gas-solid reaction, are included as the solid-state reactions.

The criterion of the *solid-state* reaction is rather ambiguous, if the microscopic insight is taken into consideration. For most of the solid-state reactions, except for a single crystal-to-single-crystal transformation, a step of molecular loosening at the reaction site should be involved earlier or later in the whole reaction process, wherein a small fraction of the molecules is not packed in an ordered crystal lattice [3]. On the other hand, most of the so-called 'solid-state decomposition' might be best considered to involve reactions that occur immediately at melting followed by instantaneous solidification due to the high melting product. These are not solid-state reactions. However, if such successive events occur in a microscopic level at the reaction front, wetting or melting of the bulk solids could not be detected during the reaction by means of thermal analyses such as differential scanning calorimetry (DSC) or of other spectroscopic methods. Then, how is it possible to distinguish such local phenomenon from molecular loosening in genuine solid-state reactions? In this chapter, so far as the bulk solid state is retained throughout the reaction in spite of the involvement of molecular loosening or microscopic melting, we regard such to be solid-state reactions and discuss them in terms of the mechanism of the solid-state reaction.

The situation noted above would be more clearly appreciated for the reaction induced by grinding. During grinding there is a release of energy and that might lead to local melting and recrystallization at the site where the reaction is taking place. Visual and/or spectroscopic observation could not identify such molecular loosening occurring at the reaction front.

There are some reactions in which the final state at the completion of the reaction turns to a molten phase, even though the initial state is a solid mixture. Needless to say, these are not included as solid-state reactions. The solid-state reactions are solvent-free reactions [4]. However, the solvent-free reactions are not always solid-state reactions.

The purpose of this chapter is to survey the general principles and concepts associated with the solid-state organic ionic reactions. The essence of the thermally induced ionic reactions is described through relevant examples classified according to the type of the reactions.

2. Factors Affecting Reactivity of Solid-State Ionic Reactions

The factors that govern the solid-state ionic reactions differ more or less from those in the long-familiar organic reactions in liquids. In this section, the characteristic factors in the solid-state ionic reactions are considered in relation to those in liquid-phase reactions.

2.1. Steric factors

Steric factors of reactions in the liquid phase are associated with the molecular structures. For the solid-state reactions these are principally related to the effect brought about by the crystal lattice. The most characteristic feature of crystalline phase reactions is the topochemical principle [5,6]. This principle is stated as that the crystalline phase reactions tend to occur with minimal atomic/molecular motion, because the molecular movement is highly restricted in the crystal lattice. Thus, the molecular packing controls the solid-state reactivity. The topochemical principle has been originally applied to photochemical dimerization of a variety of cinnamic acids at ambient temperature. For thermal reactions, in particular, under high temperature, the topochemical principle might not be necessarily applied, because the molecules restricted in the crystal lattice could be loosened thermally and released from the crystal lattice control.

Based on the topochemical concept, many concrete ideas about solid-state reactions have developed, such as the reaction cavity concept [7,8], the postulate of steric compression [9], and the role of local stress [10]. These concepts as well as the topochemical principle are concerned with the starting crystals rather than the products and invoked for the initial step of the solid-state reactions. The process initiated topochemically is considered to be followed by molecular loosening, molecular change, solid-solution formation, and then separation of the product phase [3]. If the molecular and/or crystal structures of the reactant are similar to those of the product compound, the minimum motion would hold for all the atoms in the structure to result in only a slight change of the whole framework. Such a reaction is designated as a *topotactic* reaction, originally proposed in the inorganic solid-state reactions. The topotactic reaction is at the same time a topochemical reaction. An extreme case of topotactic reaction is a single-crystal-to-single-crystal transformation with minimal disruption of the crystal lattice. Single-crystal-to-single-crystal transformations have been found increasingly in solid-state photoreactions.

On the other hand, a non-topochemical nature has been claimed in the single-solid reactions based on atomic force microscopy (AFM) or scanning tunneling microscopy (STM) observations [11,36]. The change in profiles of the crystalline surfaces during the reaction is interpreted by means of long-range migration of the molecules. The non-topochemical mechanism has been proposed also for solid-solid reactions caused by grinding and for gas-solid reactions as well. In these reactions, the topochemical principle would not be so strictly valid, because molecules are required to travel over to the other component molecules and to enter their crystal lattice.

There could be another type of non-topochemical reaction, which is initiated by molecular loosening or partial melting at the surface or defect region of the crystals.

71

Subsequently, with an increase in the number of defects caused by the initial reaction, further reaction could take place in the interior of the crystal. In such cases, the molecular identity would be mostly retained. In the following two sections typical examples of topochemical and non-topochemical reactions in the solid-state ionic reactions are described.

Polymorphism of crystals could be considered to correspond to isomerism of molecules. The discrete reactivity depending on the polymorphic forms is sometimes observed for the solid-state reaction induced under crystal-lattice control. Amide **1** has been found to crystallize in two polymorphic forms (α- and β-forms), both of which undergo thermal transformation to quinazolone compound **2** with accompanying dehydration [12]. The α-form, in which the NH hydrogen of the acetamide group is hydrogen-bonded intramolecularly with the carbonyl of $CONH_2$, is converted to the β-form, which is a direct precursor to the quinazolone product. The conversion of the α to β-form involves loss of an intramolecular NH$\cdots$O hydrogen bond and formation of a new intermolecular NH$\cdots$O hydrogen bond.

2.2. Electronic factors

The ionic reactions in solution are crucially affected by the substituent via its electron-withdrawing or -donating nature. In the case where the solid-state reactions propagate via molecular loosening, the electronic effects would be exerted as expected for solution reactions because the molecular identity is sustained. When the reaction is controlled by the crystalline lattice, the reactivity of organic compounds depends on a balance between molecular packing factors and electronic properties.

On heating at 110°C triphenylmethylalcohol derivative **3** undergoes solid-state dehydration to give quinonoid **4**. The bromine derivative and methyl derivatives are isostructural, whereas the dehydration of the dibromo derivative proceeds much faster than that of the dimethyl derivative [13]. In this case, probably, the difference in the reactivity is attributed to acidity of the phenolic hydrogen; it is higher for the dibromo derivative than the dimethyl derivative.

The Hammett equation is a useful tool for investigation of the mechanism of ionic reactions in solution. Also for the solid-state reactions, the existence of linear energy correlation has been demonstrated for the reaction rates of *p*-substituted anilines with *p*-benzoquinone to form charge-transfer complexes [14].

2.3. Medium effects

The molecules and ionic species in the crystals are stabilized by their own media instead of the solvents in solution. Thus, the solvent effect in the solid state occurs as a factor of intermolecular interaction of the molecules itself. The polarizability of the molecules should play a role in stabilization of the ionic species as in a solution. The positive and negative ions generated in the solid state locate closely together by Coulombic interactions, being stronger than van der Waals interactions. This should bring about the characteristic effect in the solid-state ionic reaction. However, closer examination focusing on this point has not been undertaken.

"Solvent effect" is a self-inconsistent concept for the solid-state reactions. However, with respect to the clathrate crystals that incorporate solvent molecules as guest species, the solvent effect in the solid state is conceivable, although the essential factors of *solvent effect* originated from the alternation of the crystal structures. The photochromic process of salicylideneaniline **5** is different for the crystals enclathrating methanol and acetone [15]. The thermal fading reaction of the photochrome is faster for acetone solvate. The photochromic behavior of **6** in its inclusion crystals depends on the guest solvents [16].

5   6

The term 'solid-state solvolysis' is somewhat questionable. However, clathrate crystals, including alcohols as a guest component, undergo nucleophilic substitution retaining the solid state, providing an example of *solid-state solvolysis* [17,18]. In *solid-state solvolysis*, it is not necessary to dissolve the substrate in liquid solvent.

A defect in crystals has been known to have an important effect, in particular, on inorganic material related to ionic conduction, doping effects in p–n junctions in semiconductors, color sensors, image development in photography, the mechanism of gas sensors, and many more. In organic solid-state reactions lattice vacancies could be a favorable site for the molecular loosening and for the formation of a new phase. Thus, when the reactions are carried out by gas–solid contacts or by grinding, the concentrations of point defects are decisive for the rates of local chemical reactions. For solid-state photoreactions the defect sites may function as an energy trap during photo-excitation [7].

2.4.   Kinetics and thermodynamics

The kinetics of the solid-state reactions are not so simple as in solution reaction. Time-conversion of the reactant or products sometimes makes it possible to estimate the apparent rate constant. For single-solid reactions mostly zero or first-order rate constant has been obtained as an apparent rate (Table 1). If the reaction starts at several nucleation centers, the rate should gradually slow down as seen in the first-order kinetics, because the reaction fronts from the adjacent nucleation centers merge or the concentration of the starting phase is decreased with reaction time. The zero-order kinetics implies the absence of such retardation. It seems likely that there are rather some acceleration factors in the later stage of the reaction. The defect multiplication would be accompanied by the progress of the reaction and plays an autocatalytic role; as the reaction proceeds, the rate is increased. In fact, solid-state reactions frequently exhibit an induction period.

The activation energy does not necessarily correspond to the activation energy in the 'true chemical reaction step' in the overall solid-state reactions. In particular, for solid-solid reactions or gas-solid reactions, not only the interface event but also diffusion, nucleation and other factors should be taken into consideration.

Let us consider an ideal solid-state reaction under the lattice control. The activity of each reactant is 1, implying $\Delta G = \Delta G°$. When the reaction occurs, $\Delta G = \Delta G° < 0$ holds and therefore ΔG is constant. The reaction proceeds completely until one component has been consumed. There is no equilibrium. The entropy change ΔS is generally a quite small value for the solid-state reactions. Therefore, because of $\Delta G = \Delta H - T \Delta S \sim \Delta H < 0$, the solid-state reactions are exothermic in nature (Table 1).

3.   Topochemical Ionic Reactions

The most well-characterized example of the topochemical ionic reaction is the rearrangement of methyl *p*-dimethylaminobenzenesulfonate (**7**) to zwitter ionic

Table 1   Heat of reaction and reaction order

Reaction		Heat (kJ mol^{-1})	Order	Ref.
			1st (X=Cl) zero (X=Me)	[20]
		23	zero	[21]

Table 1 Continued

Reaction	Heat (kJ mol^{-1})	Order	Ref.
(triazine Me-O, N-Me derivative) → (triazine Me-N, N-Me, O derivative)	133.7 total of three steps	1st	[105]
(naphthalene-COOH, N$_2^+$Br$^-$) →(100~135 °C) (naphthalene-COOH, Br)	150		[22]
Me$_2$N–C$_6$H$_4$–S(=O)$_2$–O–Me →(90 °C) Me–N(Me)–C$_6$H$_4$–SO$_3^-$	63–65		[23]
HO–C(Ph)(...)–C$_6$H$_2$X$_2$–OH →(110 °C, –H$_2$O, X=Me, Br) (Ph)$_2$C=C$_6$H$_2$X$_2$=O		zero	[13]
NH$_4^+$ OCN$^-$ →(80~90 °C) H$_2$N–CO–NH$_2$	205	zero	[24,100]
(binaphthyl NH$_2$, OH, CO$_2$Me) →(150 °C) (binaphthyl NHMe, OH, COOH)	18.6		[80]
(norbornene anhydride) →(95 °C) (norbornene anhydride rearranged)		1st	[25]

N,N,N-trimethylbenzeneaminium-4-sulfonate (**8**) on heating below the melting point of 90–91 °C [26]. The reaction is endowed with the ideal features for a topochemical process. Firstly, the molecules of the starting compound are constructed of a push-pull substituent to give a high dipole moment and, therefore, a high melting point. This structure is also advantageous for the molecule to arrange in a head-to-tail manner with the N-methyl group close to the O-methyl group to facilitate the methyl rearrangement. Secondly, the product is an ionic salt that has a high melting point (>350 °C). The zwitter ionic product is again stabilized by the alternating arrangement of the positive and negative charges.

Me$_2$N–C$_6$H$_4$–S(=O)$_2$–O–Me ⟶ Me$_3$N$^+$–C$_6$H$_4$–SO$_3^-$

7   **8**

The reaction proceeds at a considerably faster rate in the crystal than it does either in the melt or in solution by 25–40 times. The intermolecular nature of the reactions was ascertained based on 'a double label scrambling' experiment. When a sample containing equal amounts of deuterium-labeled compound D_9 (NCD_3, NCD_3, OCD_3) and a non-labeled compound was allowed to undergo solid-state rearrangement, a mixed product was formed as revealed by field desorption mass spectrometry; equal amounts of zwitter ionic products-D_0, D_3, D_6, D_9, the expected intermolecular ratio, are produced [26].

The X-ray data indicate that the molecules in the crystals are oriented such that the nitrogen is almost perfectly aligned with respect to an adjacent methyl group ($N\cdots CH_3$; 3.54 Å) [27a]. The intermolecular methyl transfer over the distance of 3.54 Å is adequate for topochemical migration. The location of empty and filled spaces and their volumes are estimated based on the molecular volumes to show there is plenty of extra space for the nitrogen atom to allow pyramidalization [27b].

A chain reaction sequence is proposed for this rearrangement. The preorientation in the crystal is directly implicated in lowering the entropy of activation by fixing the relative orientation of the reaction sites and facilitating a chain-reaction sequence (Figure 1). A theoretical study on the topochemical initiation reaction indicates the cooperative rotation of the molecules around their molecular long axes coupled with structural deformation of the methoxy group [28]. Raman phonon spectroscopy has been applied to this reaction [29]. Phonons are low-frequency intermolecular vibrations of a crystal. When the solid solution is formed and its concentration is changed, the phonon frequencies change monotonically, designated as phonon amalgamation behavior. On the other hand, in the case where no solid solution forms, the unchanged phonon bands of the two components superimpose. Experiments for the thermal conversion phonon amalgamation have not been observed, indicating that the thermal rearrangement proceeds by a heterogeneous transfer mechanism, that is, long-range cooperative rearrangement rather than a local rearrangement. Thus, the reaction has been taken as the most definitive example of an anharmonic phonon-phonon assisted reaction.

There are discrete surface and bulk processes in the solid-state organic reactions. Diffuse reflection IR, a method that monitors only the outer regions of the solid, was used to show that the surface reaction is decidedly slower than the bulk reaction [30]. Since the surface molecules are more poorly aligned at the surface, the methyl transfer rate in this region is reduced.

Organic solid-state reactions are influenced by a large change in the lattice energy even if the stability of a reactant molecule is higher than that of a product, particularly in a reaction by which molecular polarity is considerably changed. The energetic aspect of the conversion from the neutral to ionic crystals, has been discussed. The rearrangement

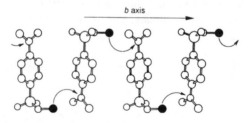

Figure 1 Schematic of the methyl migration drawn on the basis of the crystal structure of **7** (from [33] with permission from Elsevier Science).

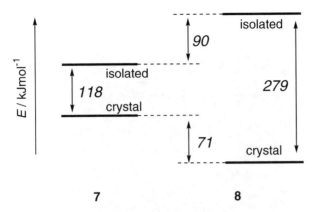

Figure 2 Energy correlation diagram of the solid-state rearrangement of **7** to **8** based on ab initio MO calculations.

of **7** to **8** is an exothermic reaction with heat of reaction of 63–65 kJ mol^{-1} as an absolute value [31]. Sarnma and Dunitz [32] carried out crystal structural analyses of **7** and **8** at several temperatures and calculated the lattice energies of both crystals. Their results indicate that there is a large difference of more than 200 kJ mol^{-1} in lattice energy between the **7** and **8** crystals. More recently, Oda and Sato [33] has carried out *ab initio* MO calculations and showed the energy correlation diagram as shown in Figure 2, in which the heat of reaction, 71 kJ mol^{-1}, is comparable with the experimental value. It is obvious that the exothermic nature of the reaction is predominantly derived from the difference in the lattice energies of the two crystals.

4. Non-Topochemical Ionic Reactions

The microscopic events in grinding two components leading to solid-solid reactions are not fully understood. One component has to move into the crystal lattice of the other component or continuously develop a fresh surface to expose to the other component.

The atomic force microscopy (AFM) technique has been employed for investigation of solid-solid reactions to gain insight into the microscopic surface phenomenon. Based on this supermicroscopic technique various examples of non-topochemical solid-state reactions have been demonstrated. The reaction induced by cogrinding amine and aldehyde is a relevant example. This solid-state reaction proceeds without melting to afford azomethine with accompanying loss of water between 80–90 °C [34].

In order to employ the AFM technique for this condensation reaction, a small crystal of *p*-chlorobenzaldehyde was placed on the surface of a single crystal of *p*-nitroaniline. After several hours the aldehyde crystal disappears completely both by reaction and by sublimation. AFM measurements disclose the formation of craters and numerous stumpy protrusions on the originally flat surface (Figure 3). This indicates that no uniform close contact is achieved between the two surfaces. Such complicated surface transformations and nonhomogenious evolution of surfaces indicate long-range molecular migrations rather than minimal movements predicted from the topochemical principle.

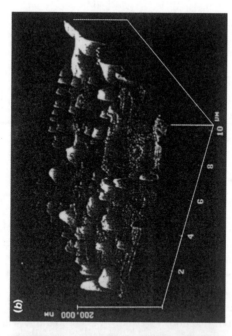

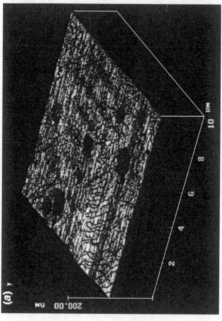

Figure 3 AFM topographies of *p*-nitroaniline on (100); (a) pure and (b) after reaction with a tiny crystal (approximately 0.3 mg) of *p*-chlorobenzaldehyde, which completely disappears after 4 h, in the previously covered area (from [34]. Reproduced by permission of the Royal Society of Chemistry).

The AFM technique is effectively used in the investigation of the reactions induced by gas-solid contact. Long-range anisotropic molecular movements occur upon chemical reaction, which moves down into the crystal. Such local phase transformation provides good chances for the reaction of a new surface by continuous disintegration of the original surface. Gas/solid reactions of nitrogen dioxide with organic substrates are investigated by means of AFM. Electron and oxygen atom transfer, nitration, and oxygenation are achieved in good yield by the reaction with NO_2 [35]. Crystalline thiohydantoin **9** reacts quantitatively with gaseous methylamine or dimethylamine with cleavage of the 3–4 bond to yield crystalline compound **10** [36].

Non-topochemical reactions are also conceivable in single-solid reaction, in which the reaction is initiated local melting or molecular loosening, as already described in Section 2. The molecular motion in the crystal lattice can occur easily in thermal reactions as compared with photoreactions, in particular, at the points of defects, which are likely to provide pockets of free spaces for molecular motion. The vinylogous pinacol rearrangement is a relevant example. On heating dihydroanthracenediol **13** over 180–240 °C, these crystals are transformed quantitatively, with accompanying dehydration, to crystals of anthrone derivative **14** without wetting or melting [21]. Similarly, compound **11** undergoes the solid-state 1,4-rearrangement to the corresponding anthrone derivative. The rearranged group is quite bulky and moves to the position of 1,4-relation, which is a rather long-distance movement accompanying large changes in the molecular and crystal structures. Temperature-dependent FT-IR of **13** indicates that the reaction proceeds with progressive change, indicating a direct transformation without melting (Figure 4). Diol **13** forms a mixed crystal (solid solution) with its isomeric compound **12** and the crystal structure of the mixed crystal is isostructural with that of **13** (Figure 5). By using this mixed crystal, the rearrangement was proved to proceed intramolecularly.

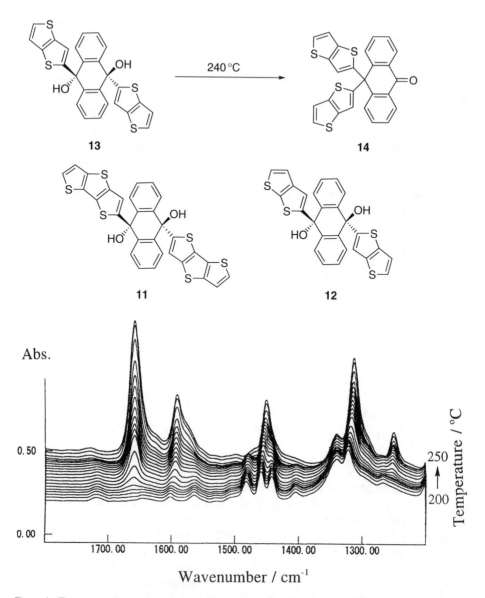

Figure 4 Temperature-dependent FTIR spectrum of **13** in the 1200–1800 cm^{-1} region, recorded by upper shift with raising the temperature from 200 to 250 °C (1 °C min^{-1}) (from [21]. Reprinted with permission. ©American Chemical Society).

It seems most likely that the reaction begins at crystal defects and/or crystal surfaces and proceeds via molecular loosening at the reaction front, followed by a unimolecular reaction and microscopic reconstruction of the crystal phase of the product. The elimination of water would be advantageous for the molecular change and reorganization, as this would leave voids in the crystal lattice to provide freedom of molecular

(a)

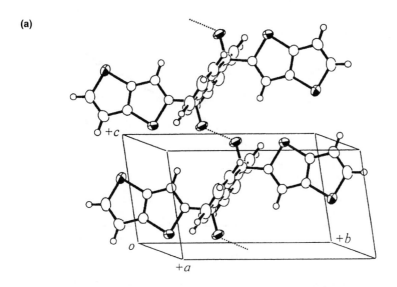

(b)

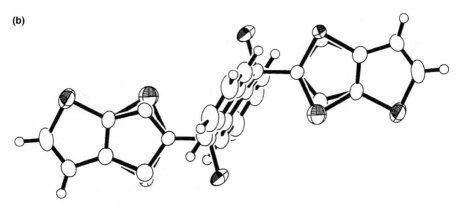

Figure 5 (a) Crystal structure of **13**. (b) ORTEP view of the disordered molecules in a mixed crystal of **13** and its isomer **12** (from [21] (p. 81)).

movement. It is interesting to note that the reaction of **13** follows zero-order kinetics (Figure 6). A possible interpretation would be based on catalytic proton transfer, wherein proton liberation and consumption take place sequentially between the neighboring molecules along the hydrogen-bonding chain (**Scheme 1**).

anthrone **14** H_2O anthrone **14** H_2O anthrone **14**

ROH ⟶ R⁺ ⟶ H⁺ ⟶ R⁺ ⟶ H⁺ ⟶ R⁺ ⟶ H⁺ ----➤

R-OH = **13**   R-OH   R-OH

Scheme 1

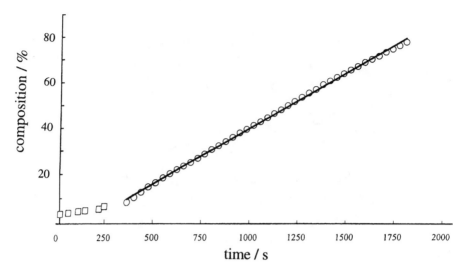

Figure 6   Time-dependent composition of **14** in the solid-state rearrangement of **13** at 200 °C, as monitored by increase of the carbonyl absorption in the FTIR spectra.

5.   Generation of Ions in the Solid-State CT Interactions

In connection with the solid-state ionic reactions a fundamental phenomenon is generation of ions in the solid state. This could be realized most primitively by charge-transfer (CT) interaction. Depending on the magnitude of electron-donor and -acceptor properties, the electric nature of CT complexes ranges from neutral to ionic. It is well known that neutral CT complexes ($D°A°$) are formed when $I_d - E_a > \alpha V$, while the crystal of $I_d - E_a < \alpha V$ affords ionic CT crystals (D^+A^-), where I_d is ionization energy of the donor molecules, E_a is the electron affinity of the acceptor molecules, V is Coulombic energy for a pair of D^+A^- and α is the Madelung constant. When the charge-transfer occurs completely to form ionic crystals, the stabilization energy is αV for each D^+A^- pair.

Crystals of neutral molecular complexes are transformed to ionic crystals by phase transition when the above-noted energy is gained. The occurrence of such phenomenon has been predicted by McConnell *et al.* [37] for the CT complexes, in which the donor and acceptor molecules are stacked alternately. This phenomenon is discovered in a CT complex of TTF and chloranil [38,39]. Not only by applying pressure (10 kbar) but also by lowering temperature below 80 K under the ambient pressure, the complex shows a

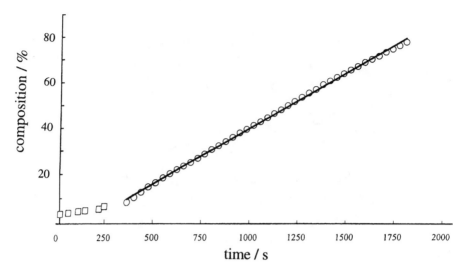

TTF   perylene   bromanil   TCNQ   DDQ

neutral-to-ionic transition with accompanying color change. Under 80 K the magnitude of electron transfer increases, principally due to the decrease of cell constant, hence the increase of the Madelung energy stabilizing the ionic state.

Charge-transfer complexation is accomplished not only by recrystallization of the donor and acceptor components from solutions, but also by cogrinding the solid-donor acceptor compounds [40,41]. The solids obtained by cogrinding tetrathiafulvalene (TTF) and 7,7,8,8-tetracyanoquinodimethane (TCNQ) exhibit high electrical conductivity of similar magnitude as found in the TTF-TCNQ complex obtained by recrystallization [42]. It seems likely that the segregated columnar structure of TTF and TCNQ and partial charge-transfer as well, both being essential requirements for organic conductors, have been realized also by cogrinding TTF and TCNQ. Cogrinding is effective for the formation of hydrogen-bonded supramolecular assemblies [43,44].

By grinding a mixture of donors and acceptors in the solid state, imperfections containing unpaired electron spins can be introduced. Thus, the electron spin resonance (ESR) signal is observed in the powder sample obtained by grinding a mixture of perylene and bromanil [45]. A similar phenomenon has been observed in non-planar compounds. Upon cogrinding with DDQ or TCNQ in a mortar with a pestle, compound **15** gives rise to amorphous solids which involve radical ionic species, as probed by UV/Vis and ESR spectra [18,19]. For molecules with irregular shape such as **15**, periodic packing to form crystals of CT complexes would be unfavorable and rather takes a random orientation in the coground solids. Even in such solids, charge transfer occurs fractionally at the interface of the donor and acceptor to allow location of unpaired electrons.

It should be noted that the 1:1 CT complex obtained by cautious recrystallization of compound **15** together with DDQ or TCNQ shows no ESR signal, while the signal appears and is enhanced by grinding its crystals [46]. The results indicate that even in homogeneous solids of CT complexes grinding increases unpaired electron spins.

$$\textbf{15} \bullet \text{DDQ} \xrightarrow{\text{grinding}} \textbf{15}^{+\cdot} \ \ \text{DDQ}^{-\cdot}$$

1:1 CT complex

The above observation is in line with the following experiments reported by Matsunaga [47]. When *N*-picrylanilines are melted and then quickly cooled, the resulting super-cooled melt shows a deeper color than in the crystalline state, indicating that charge-transfer interactions are enhanced. For example, *N*-picryl-*m*-anisidine is yellow in the crystalline state, while it is reddish-orange in the molten and super-cooled state. Even when the molecular packing in the crystalline state is unfavorable for the intermolecular CT interactions, a contact between the two moieties may be allowed by the disorder introduced by melting.

Photoexcited electron transfer in the crystalline molecular complexes generates an ionic species in the solid state. When cocrystals of benzyl cyanides with tetracyanobenzene are irradiated, intermolecular reaction occurs to afford a stilbene derivative. This reaction is assumed to proceed via CT and eventually to give a radical pair [48]. Another interesting photoreaction via photo-excited electron transfer is concerned with asymmetric synthesis. Bisthiadiazolotetracyanoquinodimethane **16** forms weak charge-transfer complexes with *o*-divinylbenzens in a 1:1 composition. Upon charge-transfer excitation of these crystals, the complex affords adduct **17** via single-crystal-to-single-crystal transformation [49]. Furthermore, the adduct is obtained in an optically active form with 95% enantiomeric excess. This reaction provides an example of absolute asymmetric synthesis, because the complexes crystallize in an asymmetric packing ($P2_1$) although both components are achiral compounds.

This type of 'absolute asymmetric synthesis' based on spontaneous asymmetric crystallization of achiral compounds coupled with their topochemical transformation has been well known in solid-state photoreactions. Two-component crystals derived from acridine (**18**) and diphenylacetic acid are both achiral compounds, while these form a crystalline molecular complex with a chiral space group ($P2_12_12_1$) [50]. Upon irradiating the complex, photosensitized electron transfer occurs from acridine to the carboxylic acid and subsequent hydride transfer followed by decarboxylation gives rise to a hydroacridine radical and a diphenylmethyl radical. The next radical coupling occurs with the shortest distance of 5.1 Å between the two preradical carbon atoms in the crystal lattice to afford (*S*) (–)-**19** as the major enantiomer.

Two-component molecular crystals are used for photo-induced electron transfer to give a radial pair in a series of cocrystals of 2-thienylacetic acid with aza-aromatic compounds [51].

Acenaphthylene (**20**) and TCNE form a 1:1 CT complex. The CT excitation of this complex gave a [2+2]-adduct **21** in contrast to photochemical inertness in solutions.

Irradiation of the powder of an equimolar mixture of **20** and TCNE affords not only adduct **21** but also a small amount of *cisoid* and *transoid* dimers of acenaphthylene [52]. The formation of these dimers indicates that a part of radical cation **20**⁺· produced by CT excitation can add to a ground state **20** to give (**20**)₂⁺· radical cation, which is assumed to be the precursor of the dimer of **20**. A looser and more random arrangement of the molecules of **20** and TCNE in the mixed powder should allow the dimerization of **20**.

6. Acid and Base in the Solid-State Reactions

The reaction between an acid and a base leads to the formation of a salt, which is an ionic material and hence has sufficiently high melting point to be of merit in thermal reactions. By the use of gaseous acids or bases such as hydrogen chloride and amines, the acid-base reaction can be carried out at ambient temperature. Paul and Curtin [53] investigated the reaction of ammonia gas with carboxylic acid and its derivatives and demonstrated anisotropic properties of the gas-crystal reactions. Single crystals of (*R*)(+)-2,2-diphenyl-1-methylcyclopropanecarboxylic acid (**22**) react readily with ammonia gas to give the 1:1 ammonium salt [54]. The reaction proceeds almost exclusively in the direction parallel to the *b*-axis. The X-ray structure shows that the carboxyl-carboxyl hydrogen-bonded chains runs parallel to the *b*-axis and the chain is surrounded by an assembly of the hydrocarbon moieties [54]. For such molecular packing, attack of ammonia molecules to a carboxyl group would be allowed only from the direction parallel to the *b*-axis. The reaction, which occurs preferentially from one side, as cited above, is classified as a unitropic reaction.

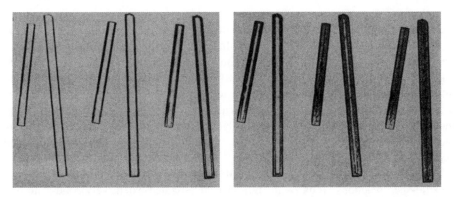

22   **23**

A unitropic reaction has also been found in the reaction of crystalline *p*-bromobenzoic acid anhydride with ammonia gas [55], which proceeds in only one direction along the

Reaction of crystals of *p*-chlorobenzoic anhydride with ammonia. The larger crystal is 1.1 × 0.07 mm. Times are from left and proceeding from left to right: 0, 57 min, 195 min; 300 min, 540 min, 24 hr [from ref. 53(b)].

polar axis (space group *C*2). In the reaction of *p*-bromobenzoic acid with gaseous ammonia the attack of ammonia in a direction normal to the top face of the crystal is unfavorable, because the aromatic ring moiety is oriented to protect against attack [56]. On the other hand, either of the two side faces of the crystal are exposed hydrogen-bonded carboxyl units available for reaction. In crystals of several benzoic acids, such anisotropic reactivities are noted. This type of crystal reactivity is referred to as ditropic.

An acid–base reaction induced by solid-solid contact has also been investigated in relation to the crystal structures. Crystals of *p*-aminosalicyclic acid hydrochloride (**24**) lose HCl when placed in contact with Na_2CO_3 at room temperature and 60 °C. This reaction always starts at the point of contact to the base and proceeds anisotropically through the crystal, resulting in dehydrochlorination and formation of the *p*-aminosalycylate anion (**25**) [57]. The anisotropic behavior is well interpreted on the basis of the molecular orientation in the crystal lattice.

The reaction of 1,2,3-trihydroxybenzene (**27**) with 8-hydroxyquinoline (**26**) is regarded as a solid state acid-base reaction, forming an organic ionic salt [58]. When solid powders of **27** and **26** are mixed thoroughly in an agate mortar, an orange-colored molecular complex is formed. Based on the X-ray analysis it is found that the proton of the middle OH group is transferred to the N atom of **26**. The color change is interpreted by intervention of charge-transfer interactions.

Acids are employed as catalysts in the organic solid-state reactions. These include inorganic acids such as heteropolyacid, zeolite, montmorillonite, metal peroxide, and so on. Hetero-polyacid, $Cs_{2.5}H_{0.5}PW_{12}O_{40}$ and polymer resins Nifron/SiO_2 have been reported to promote pinacol rearrangement by solid-state grinding [59]. Powder X-ray diffraction measurements indicate that the reactant crystals become nearly amorphous by grinding, while the crystallinity of the catalyst is unchanged. The catalytic action is thought to occur at the surface of the catalyst, and the products separate from the surface and diffuse into the organic solid phase without formation of large and less mobile crystalline particles.

As an organic acid catalysis *p*-toluenesulfonic acid (*p*-TsOH) is most frequently used in solid-state reactions. Toda *et al.* [60] have investigated extensively solid-state reactions catalyzed by *p*-toluenesulfonic acid, which include dehydration, ether formation, and pinacol rearrangement. The pinacol rearrangement of benzopinacol to give the corresponding pinacolone was investigated by means of AFM measurements to indicate that no reaction occurs on (0 0 1) of the pinacol, whereas (1 0 0) exhibits distance-dependent craters and volcano-like mounts that extend from the contact edge of the crystals and undergo phase rebuilding while reacting [61]. On (1 0 0) the hydroxy hydrogens point down, that is, the lone pair electrons of the oxygen atom are oriented to make protonation possible. It seems most likely that catalytic protons migrate from one molecular compartment to the next by proton consumption at its inside and proton liberation at its outside, which is the inside of the next compartment. Such propagation of the reaction is similar to that assumed in the thermal 1,4-vinylogous rearrangement of **13** base on catalytic action of a carbocation salt (**Scheme 1**) [62].

Two components in the crystals interact with the external gaseous compound and enter into the three-component reaction in the solid state. As described in Section 8, a host component of clathrate crystals undergoes the nucleophlic substitution by guest ethanol upon contact with HCl gas which acts as a catalyst [18]. The formation of benzodiazepine **29** from dihydrohalide of *o*-phenylenediamine (**28**) is another example of this type of gas-solid reaction, [63] wherein the acid catalyst is incorporated in advance in the crystal and acetone enters into the reaction as a gaseous external reactant.

Base-catalyzed solid-state reaction has been found in the benzilic acid rearrangement. Finely powdered benzil and KOH are heated at 80 °C for 2 h, and benzilic acid is obtained in 90% yield [64].

7. Proton Transfer

A proton is the simplest ionic species and the basis of acid-base chemistry. The factor associated with the solid-state proton transfer is not only the strength of the acid and base but also seems to depend on various factors including the crystal structures.

Benzoquinone and methylhydroquinone undergo rapid isomerization in solution to form hydroquinone and methylbenzoquinone. However, in the solid state, proton transfer is not involved; solid-state cogrinding of the two components leads completely to the unsymmetrical quinhydrone of the starting materials [65]. In the case where it is possible to prepare the complex by crystallization from solution, its color, spectroscopic properties and crystal structures are identical with those of the complex obtained by grinding the solid components together. This finding indicates that the molecules have sufficient mobility in the solid state, reorganizing the inter-homomolecular interactions to form inter-heteromolecular complexes [40–44].

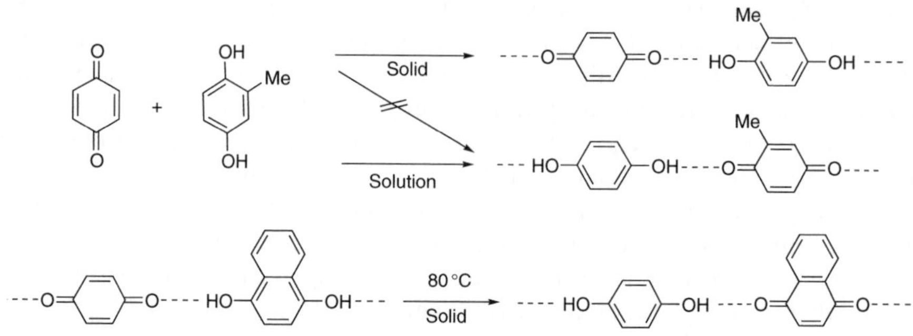

It is possible to promote proton transfer thermally in the solid state. The 1:1 complex of benzoquinone and naphthohydroquinone, which can be prepared by solid-state cogrinding, is converted to the complex of naphthoquinone and hydroquinone when heated at 80 °C [66]. The resulting solids are microcrystalline and have the same crystal structures as those prepared directly by crystallization from solution.

Single crystals of 1,5-dihalo-2,6-naphthoquinhydrone provide a unique cooperative proton-electron transfer system, in which a charge transfer state is accompanied by hydrogen transfer. The final proton-electron-cooperative transfer (PET) state is characterized as a molecular crystal of hydrogen-bonded neutral radicals (**Scheme 2**) [67].

PET phenomenon is observed as a phase transition of the single crystals by means of pressure dependence of the O–H stretching vibration in IR spectra and the CT transition energies in electronic spectra. The crystal structure of 1,5-dihalo-2,6-naphthoquin-hydrone is constituted of the alternately stacked quinone and hydroquinone components and the existence of two-dimensional hydrogen-bonded charge-transfer networks, which facilitate the intermolecular cooperative proton-electron transfer.

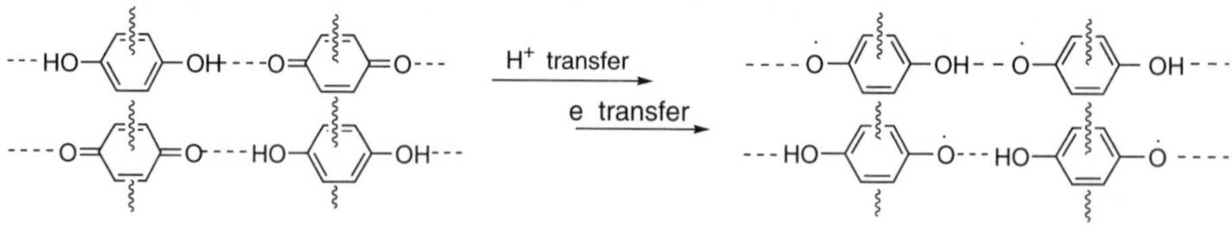

Scheme 2 Schematic representation of proton–electron transfer (PET) in quinhydrone.

Thermally induced reversible single-crystal-to-single-crystal phase change, associated with a proton transfer and a concomitant change in color, is observed in the 1:1 molecular complexes of 4,4-bipyridine (**30**) and squaric acid (**31**) [68]. These compounds form two polymorphs (monoclinic and triclinic forms), both of which are colored and consist of infinite chains connected by strong hydrogen bonds of the type NH$\cdots$O$^-$ and N$\cdots$HO. On heating to 180 °C the monoclinic form changes color from yellow to red. The maximum absorption band at 390 nm in the red form is ascribed to the CT band. This phase transition always originates at the end of the crystal and moves along the length of the crystal. The deuterated form of the monoclinic crystals, in which the acidic protons are replaced by deuterium, showed no color change on heating to the decomposition temperature. Such a strong isotope effect seems to suggest that the red high-temperature modification is formed from the yellow room-temperature modification by proton transfer. It is remarkable that no color change was observed for the triclinic form.

30   **31**

The crystals of 2-iodoaniline picrate have three polymorphic forms. The yellow needle crystals of form I (monoclinic) changed to red crystals upon heating to above 60 °C, which is ascribed to the proton transfer from ionic structure to CT form [69]. The form II (triclinic), the dark-green plate crystals, also affords the red crystals upon heating to above 60 °C. The resulting red crystals are the same as those obtained by heating the form I. Similarly, the molecular complex of picric acid and 1-bromo-2-aminonaphthalene shows a thermal phase transition [70]. The change from yellow to red is ascribed to the proton-transfer and charge-transfer.

Intramolecular proton transfer is responsible for tautomerism. Proton tautomerism such as keto-enol or amino-imino transformations is in equilibrium in solution. The feasibility of proton transfer in crystals depends on the potential profile and on the height of the potential barrier along the reaction coordinate in the crystalline environment. When the potential barrier of the proton transfer is low, proton tautomerism is responsible for dynamic disorder along the hydrogen bond between two tautomeric forms [72]. When the shape of the potential profiles along the reaction coordinate is symmetrical with regard to the saddle point, quantum-mechanical tunneling can significantly reduce the energy required and fast conversion between two equivalent forms occurs in the solid state [71]. The dynamic behaviors of the tautomeric system have been extensively investigated mainly by the use of ^{13}C CPMAS NMR.

N-salicylideneaniline and its derivatives have been known to exhibit photochromic behavior in the solid state. The photochromism is produced by photoexcited intramolecular proton transfer to give a *cis*-keto form and subsequent isomerization to a *trans*-keto form. The photo-induced form, *trans*-keto, generally reverts thermally to the original form via the ground-state proton transfer. The crystals of some other salicylideneaniline derivatives are thermochromic. It is usually the case that photochromic crystals do not exhibit thermochromism, and vice versa. The difference in this chromic behavior is ascribed to the molecular conformation and packing in the crystal. Photochromic salicylideneanilines are packed rather loosely with a non-planar structure, while thermochromic salicylideneanilines are packed tightly with a planar structure, which facilitates the proton transfer in the ground state [73].

| enol | *cis*-keto | *trans*-keto |

In contrast to *N,N'*-disalicylidene-*p*-phenylenediamine (**32**), its pyrene modification **33** shows no themochromic behavior, which is interpreted by its non-planar structure in the crystal lattice [74].

32   **33**

Crystal structural change in thermochromism of *N*-(5-chlro-2-hydroxybenzylidene)-4-hydroxyaniline (**34**) has been carried out with high accuracy to reveal that there is an equilibrium between the OH and NH forms in crystals and that the population of the NH form increases with lowering the temperature [75]. At 90 K approximately 90% of this compound exists as the NH form, which has rather a zwitter ionic character. The stabilization of the NH form is primarily attributed to the intermolecular hydrogen bonding associated with the phenolic-hydroxy group.

34   OH form   **34**   NH form

The photochromic nature of salicylideneanilines depends strongly on the crystalline environments, since the molecular motion associated with the photochromic process is large. Thus, non-photochromic salicylideneanilines exhibit photochromism when incorporated in deoxycholic acid as clathrate crystals [76]. The host acid provides sufficient space for the photoisomerization of the guest salicylideneanilines. The bulky substituents introduced in salicylideneaniline molecules act as a space-opener to maintain room for the photoinduced molecular motion in the crystal lattice [77].

The difference in the proton transfer properties of the polymorphs of 6-(2,4-dinitrobenzyl)-2,2-bipyridine (**35**) was investigated [78]. The monoclininc crystal (I) is photochromic, whereas the orthorhombic crystal (II) does not change its color upon illumination. The intramolecular CH–NO distances are 2.409 and 2.379 Å for I and II respectively, there being no difference. In I crystals the pyridine rings are π-stacked with the neighboring pyridine rings with the center–center distance of 3.77 Å, whereas in nonphotochromic II 2,4-dinitrobenzyl chromophore is tightly sandwiched between the pyridine rings, allowing strong electronic coupling between the excited chromophore and the π-system of the neighboring molecules.

35

A unique tautomer and its tautomerization via proton transfer is observed in 3-methyl-4-nitropyrazole (**36**) and its tautomer [79]. Two forms, A and B, have been obtained by recrystallization from chlorofolm-cyclohexane and ethanol-water, respectively. They have quite different crystal structures from each other. Tautomer B is stable, while tautomer A is transformed into B on standing for 6 days. The ^{13}C CPMAS spectra show the presence of both tautomers in the crystals during the transformation duration. The tautomerization proceeds via double proton transfer in the dimeric unit in the crystal of A. The occurrence of two tautomers in two different crystals, as shown in this example, is quite rare. The phenomenon is designated as desmotropy.

36 A   **36 B**

8. Substitution Reactions

A variety of substitution reactions via ionic mechanism, including S_N2, S_N1, electrophilic and nucleophilic aromatic substitutions, have been recognized so far. In order to induce the substitution reactions, the two compounds should be allowed to contact together, which would be achieved by solid-solid or gas-solid contact or by preorganization of two components in the crystals.

Even single-component crystals can be used to promote substitution reactions, if it occurs intramolecularly or inter-homomolecularly. These cases can be regarded also as rearrangements. An example has been described in Section 3 for methyl p-dimethyl-aminobenzenesulfonate (**17**), in which the methyl group attached to the sulfonate group undergoes an inter-homomolecular nucleophilic attack by the neighboring amino group [26]. Another interesting inter-homomolecular substitution has been reported for ester **37** [80]. Upon heating the racemic crystals of ester **37** at 150 °C for 15 min, conversion to carboxylic acid **38** takes place almost quantitatively (93–95%).

The intramolecular process is not likely, judging from topochemical consideration based on the X-ray crystal structure. There is a centrosymmetric (R)-(S) pair in the unit cell. However, the arrangement of this pair is again topochemically unfavorable for the reaction. The reaction should occur between the homochiral molecules, i.e., (R)-(R) and (S)-(S). This enantioselectivity was proved by double-isotope labeling, with CD_3 in one enentiomer and ^{18}O in the other. The racemate obtained by cocrystallization of the CD_3ester (R)-(+)-**37** and the ^{18}O ester (S)-(–)-**37** was thermally converted into the racemic **38**. Mass spectrometric analysis confirmed the enantioselective process; the molecules of the product are labeled either by CD_3 or by ^{18}O; no species containing both labels is detected. Thus, the reaction occurs between the homochiral molecules, i.e, within the (R)···(R)···(R) and (S)···(S)··· series. In this case, the angle of approach (N-Me-O) is 173.2°, close to 180°, providing direct evidence for the linear trajectory in the S_N2-type transition state.

The sp^3 carbon atom in aryl-substituted tert- and sec-alcohols shows high reactivity in solid-state reactions. These alcohols undergo chlorination by exposure of the powdered sample to HCl gas [60].

$R_1 = Ph, R_2 = H$
$R_1 = Ph, R_2 = Ph$

A mixture of powdered diarylmethanol and an equimolar amount of p-toluenesul-fonic acid is kept at room temperature for 10 min to give the corresponding ether in 95% yield via intermolecular dehydration [60]. Unsymmetrical ethers are prepared by solid-state reaction of cocrystals. For example, when the 1:1 cocrystals of diphenyl-methanol and di(p-tolyl)methanol, obtained by recrystallization of equimolar amounts of these alcohols, are ground with *p*-toluenesulfonic acid, the cross-coupled ether is produced as a major product (69%), while the symmetrical ether is below 3% [81].

Replacement of a thiophene ring in triphenylarylmethanols is quite effective for the generation of the stable carbocation in the solid state by the aid of an acid catalyst. The sp^3 carbon in **39** readily generates a stable carbocation species in the solid state. Furthermore, compound **39** has the molecular bulkiness and rigidity that are required for the host compound of crystalline inclusion compounds and, in fact, affords a variety of solvated crystals [82]. Thus, when **39** is recrystallized from ethanol, crystalline complexes with a host:guest ratio of 1:2 are obtained. These (**39**)(EtOH)$_2$ crystals are exposed to HCl gas at room temperature for 3 h. After working up with a dilute aqueous NaOH solution, the product analysis shows that the corresponding ethyl ether is formed in 38% yield [18]. Throughout the exposure procedure there is no sign of wetting or melting. The reaction should be designated as 'solid-state solvolysis'.

39   1:2 inclusion complex

Aromatic electrophilic substitution is achieved by solid-state cogrinding the aromatic compounds and electrophiles. When 3,5-dimethylphenol is ground with 1 mol equivalent of *N*-bromosuccinimide (NBS) in the solid state, tribromoderivative is obtained in 45% yield [83]. It is also shown that NBS reacts in the solid state readily with benzaldehydes bearing electron-donating substituents to afford bromine-substituted products with high regioselectivity [84].

40   **41**   **42**

Aromatic nucleophilic substitution is induced in a two-component molecular complex. Hydrogen-bonded cocrystals of 4-chloro-3,5-dinitrobenzoic acid (**40**) and *p*-aminobenzoic acid (**41**) undergo solid-state substitution upon heating at 180 °C, affording diarylamine derivative **42** in very high yield [85]. The substitution requires a high temperature; therefore, large molecular motions including diffusion as well as reorientation should be involved.

Bis(2-nitro-4-chlorobenzenediazonium)tetrachlorozincate (**43**) decomposes gradually, and a several year-old sample of **43** has an altered composition [86]. It is converted, retaining the solid state, to 2,4-dichloro derivative **44**, where anionic species X^- is of undetermined anionic composition. When the solids of **43** are subjected to elevated temperature (84 °C for 41 h) the same transformation is induced. This conversion is regarded as a solid-state aromatic nucleophilic substitution reaction; the nitro substituent is replaced by chlorine. Diazonium salt **45** decomposes in the solid state with evolution of nitrogen to afford 3-bromonaphthalene-2-carboxylic acid (**46**) [87].

9. Addition and Elimination Reactions

Electrophilic additions in the solid state have been carried out mostly as gas solid reactions. Upon exposure to bromine vapor, solid α, β-unsaturated acids yield *trans*-adduct in high yields, where addition of bromine in solution is difficult [88]. The *trans* adduct seems to indicate the involvement of the bromonium ion as an intermediate even in the solid state. The stable bromonium ion of the hindered olefine, adamantantylideneadamantane, has been prepared and their X-ray crystal structures have been determined [89].

Gaseous bromine reacts with chiral crystal of 4,4'-dimethylchalcone ($P2_12_12_1$) to afford chiral dibromide [90]. The crystal structure of the chalcone shows marked nonplanarity, which differentiates between the front- and rear-sides of the molecule. The tilted carbonyl and phenyl groups block the access to the double bond from one side, leading to the observed optically active product [91]. The reaction of bromine vapor with chiral crystals ($C2$) of **47** results in the formation of the rearranged product, **48**, in up to 8% enantiomeric excess [92].

Asymmetric induction in the addition reaction is achieved by use of the chiral environment provided by cyclodextrin. Crystalline complexes of cyclodextrine with unsaturated carboxylic acids such as crotonic, methacrylic, maleic acids undergo asymmetric hydrobromination and halogenation in gas–solid reactions [93]. For example, (–)-2,3-dichloro-2-methylpropanoic acid is isolated in 100% optical yield when an α-cyclodextrin complex of methacrylic acid is chlorinated.

Intramolecular Michael-type addition reactions are observed in chalcone **49**, which yield the corresponding flavanone **50** in the solid state at temperatures 50–60 °C, significantly below the melting points of the reaction products [94]. The X-ray crystal structure reveals that the phenolic OH is intramolecularly hydrogen-bonded to the carbonyl group. In order to lead to the observed product, bond rotation is required to bring the phenolic oxygen atom close to the β-carbon atom, indicating that the addition most likely proceeds in a non-topochemical fashion. Reactions may occur first at the surface of crystals or defect region where the molecules are free to rotate. Subsequently, with an increase in the number of defects caused by the initial reaction, further reaction could take place in the interior of the crystal.

49 $50-60\,°C$ **50**   X=Cl, Br

A 'vibrating mill' can activate the reaction system by bringing the reagents into very close contact and by providing extra mechanical energy. By means of this technique, dumbbell-shaped C_{120}, [2+2]-dimer of fullerene (**51**), was synthesized [95]. Vigorous vibration of a mixture of C_{60} and excess KCN powder for 30 min under nitrogen leads to the formation of 18% of C_{120}. A mechanism similar to that of the so-called benzoin condensation is assumed. A cyanide ion plays a role as a catalyst via the first nucleophilic attack to C_{60}, assisting the 1,4-type addition to another C_{60} moiety. This technique was also used for preparation of the C_{60}-C_{70} cross-dimer C_{130} [96]. In this case, 4-aminopyridine works as a catalyst.

51

Among elimination reactions induced in the solid state, dehydration is the most frequently encountered. Powdered 1,1-diphenylpropan-1-ol kept in a dessicator filled with HCl gas for 5.5 h gives pure 1,1-diphenyl-1-propene in 99% yield. The dehydration reaction proceeds much faster by using Cl_3CCOOH as a catalyst [60]. The dehydration reaction in a single-solid of 3,5-disubstituted-4-hydroxyphenyldiphenylmethanol (**3**) shows some preference for reaction in the direction of the hydrogen-bonded chains

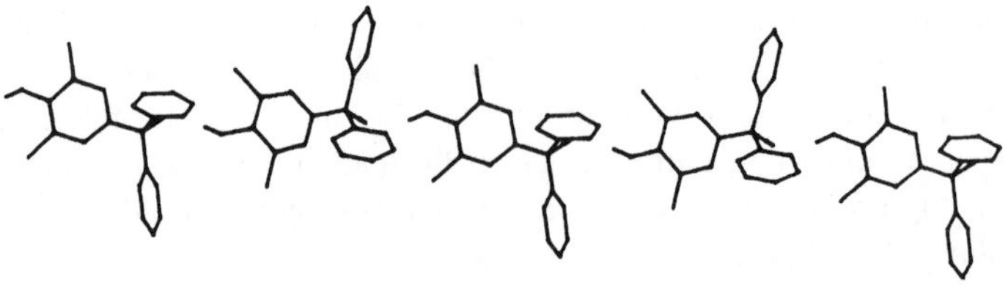

Figure 7 Crystal structure of **3** (X=CH₃), showing the hydrogen-bonded chain along the *c*-axis (from [13]).

(Figure 7). The compounds crystallize with the phenolic hydroxy group prealigned by hydrogen bonding to the alcoholic hydroxyl group of an adjacent molecule in a fashion suitable for self-catalyzed dehydration reactions [12]. It is interesting to note that photo-irradiation to the crystal can serve to produce nucleation sites for the thermal reaction. The exposed portion of the crystal clearly showed a much greater thermal reactivity than that unexposed [13]. Probably, the photoinduced deprotonation of the phenolic hydroxy group takes place locally in the bulk crystals, and the proton acts as catalyst to initiate the thermal dehydration.

A similar dehydration reaction is induced in a naphthyl analog **52** thermally at 80–90 °C to give a quinonoid compound, whereas the corresponding thermal reaction to give a quinonoid compound is not observed in **53** [97]. The difference in these reactivities is attributed to the molecular conformation: the central sp³ carbon atom lies from the plane defined by the three bonded carbon atoms by a distance of 0.421 Å in the naphthalene derivatives, while in the phenol derivative **53** it is 0.463 Å. The greater tendency toward planarity in the naphthalene compound may be a factor in the greater solid-state reactivity.

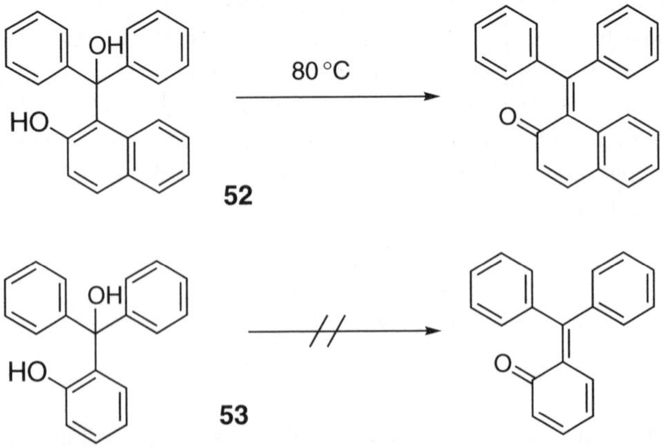

Indandione compound **54** has been found to undergo thermal dehydration in the solid state to indantrione hydrazone **55** [98]. The reaction is slow; after 3.5 weeks at 65 °C the conversion is 24%. For a single crystal, the dehydration shows anisotropic behavior. The preferred direction of the reaction is that along the hydrogen-bonded chain.

54   **55**

Stereospecific elimination reaction via gas-solid contact is realized. The *meso* form of dichloroadipate gives *trans, trans*-hexadienedioic acid ester. This is explained by the *trans* elimination of HCl in an antiperiplanar conformation in the crystals [99].

10.   Rearrangement

The first example of solid-state chemical reactions, the transformation of ammonium cyanate into urea, has recently been revisited [100]. The crystal structure of ammonium cyanate indicates the presence of NH···O hydrogen bonding for each of the hydrogen atoms in the NH_4^+ ion. For the rearrangement to occur, the proton transfer from NH_4^+ to the nitrogen atom of NCO^- should be required at the onset of the reaction, suggesting that the orientational loosening of the cation is involved in the initial step. The rate of the reaction is enhanced suddenly on melting, which indicates that the solid-state reaction is not topochemical.

The thermal rearrangement of *O*-alkylimino ether into *N*-alkylamide, Chapman-like rearrangement, occurs intermolecularly [101]. Chapman-like rearrangement in solution occurs normally above 200 °C. In the solid state, however, the rearrangement takes place unusually rapidly, which can be rationalized based on a very favorable geometry observed in the crystal packing [102]. An X-ray crystal structure of 5-methoxy-2-aryl-1,3,4-oxadiazole (**56**) shows that the C, N atoms are effectively aligned for the conversion to **57** with a C···N distance of 2.9 Å (Figure 8). The reaction is catalyzed both by the cation B^+ and anion A^-. Once the cation B^+ and anion A^- are formed initially in any of two neighboring molecules, the transfer of a methyl group between each of these ions and an adjacent molecule along the molecular chain occurs more easily than between the neutral molecules (**Scheme 3**). The transition states of these two ionic paths have identical geometry. Hence the rearrangement is domino-like propagated in two opposite directions, since at each step the same ions are formed. Furthermore, between the cationic and anionic domino-alignments, cooperative propagation should be involved. This is an interesting mechanism, being characteristic in a crystalline state ionic reaction, and should be designated as a double ionic propagation.

Methyl cyanurates undergo thermal methyl migration in the solid state [103]. The cooperative mechanism applies also in these rearrangements [104]. 4,6-Dimethoxy-3-methyl-dihydrotriazine-2-one (**58**) undergoes the solid-state methyl migration

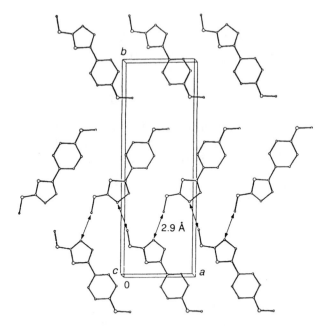

Figure 8 Crystal structure of **56** projected down the *c*-axis [from ref. 102. Reproduced by permission of the Royal Society of Chemistry].

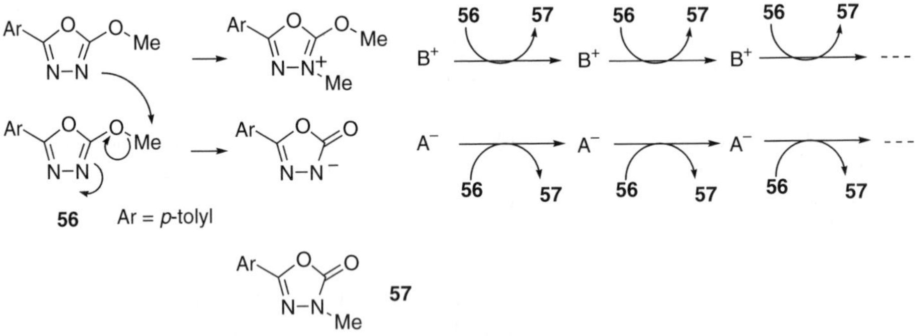

Scheme 3

intermolecularly from oxygen to nitrogen atom to give **60** [105]. The X-ray analysis reveals that the crystal structure of **58** is made up of layers consisting of planar molecules. The coplanar molecules within a layer form an infinite ribbon. The existence of the two courses of the methyl migration has been ascertained based on a quantitative analysis of the samples heated to different temperatures. The homo-chain mechanism gives **59** as a transient compound and the hetero-ribbon mechanism leads directly to the formation of **60**. The hetero-ribbon mechanism is described by a cooperative transfer of two methyl groups within the layer, as schematically illustrated in **Scheme 4**.

Methyl group rearrangement in the solid state is found in isoxazole derivative **61** [106]. This compound undergoes topochemically controlled thermal intermolecular rearrangement in the solid state to give the corresponding *N*-methylisoxazole **62** as major products. The intermolecular nature of the rearrangement was proved by a cross-experiment where equimolar amounts of the appropriately labeled compound were heated together. The efficient intermolecular migration of the methyl group is ascribed to the strong polarization of the exocyclic *O*-alkyl bond.

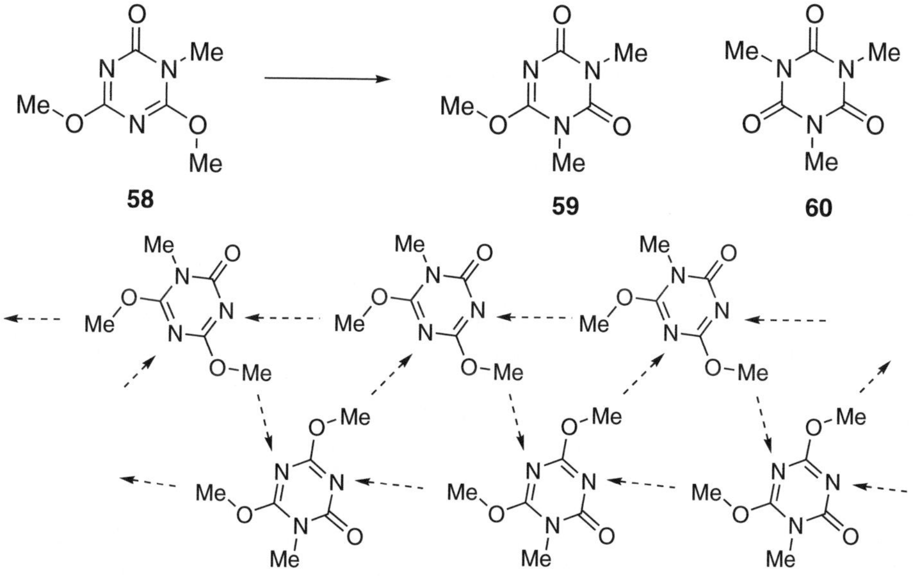

58   **59**   **60**

Scheme 4   Hetero-ribbon rearrangement.

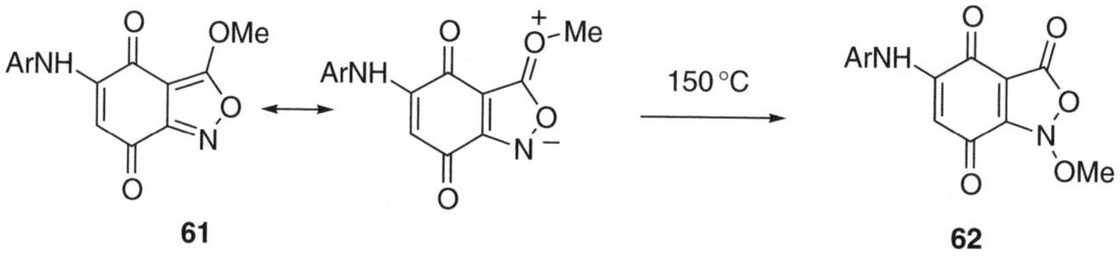

61   **62**

O-Acylsalycylamides (**63**) undergo the solid-state acyl-migration from O to N on heating around 100 °C [107]. The large mean square displacement amplitude between the amide nitrogen and acyl carbon atoms and the significant liberation of the acyloxy group were revealed by means of analysis of thermal motion based on application of the rigid-bond criterion [108]. These results are consistent with the intramolecular mechanism.

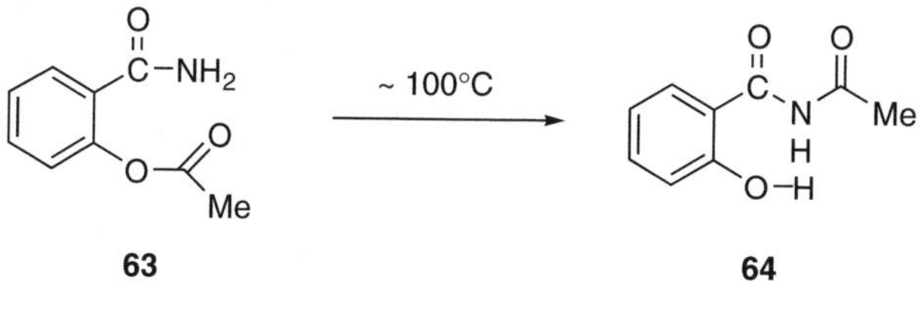

63   **64**

Solid-state grinding with acid catalyst can induce rearrangements. Propargyl alcohol undergoes TsOH-catalyzed rearrangement known as Meyer–Schuster rearrangement. The reaction is carried out by keeping a mixture of the powdered alcohol and an equimolar amount of TsOH at 50 °C for 2–3 h [60]. The pinacol rearrangement is induced in the solid state faster and more selectively than that in solution by cogrinding with p-toluenesulfonic acid [109].

The vinylogous pinacol rearrangement, which is induced thermally in single solid of **13**, occurs at room temperature to give **14**, along with the *cis-trans* isomerization, by cogrinding of **13** with an equimolar amount of p-toluenesulfonic acid using a pestle and mortar [21]. The intramolecular nature of the rearrangement is again ascertained by the use of the mixed crystals. The powder X-ray diffraction analysis discloses that the ground solids are in an amorphous state, in contrast to the solids resulting from the thermally induced rearrangement, indicating that the process is non-topochemical.

The time courses of the product distribution from *trans*-**13** and cis-**13** show significant features, namely, that the *cis* is transiently formed when the *trans* is ground (Figure 9): the *trans* isomer undergoes fast isomerization to the *cis* isomer and the anthrone **14** is formed consecutively via the *cis* isomer. These observations support the π-participation or involvement of the thiophene analogue of the phenonium ion, since the π-electrons of the thienothiophene ring in the *cis* isomer are transannularly well placed to interact with the developing carbocation (Figure 10). For the *trans* isomer, the generation of the carbocation and its return to the original configuration are both kinetically less favorable than those for the *cis* isomer. Therefore, the isomerization to the *cis* isomer is enhanced at an early stage of the reaction, but finally the 1,4-rearrangement accompanied with dehydration splits irreversibly from the common carbocation intermediate under thermodynamic control. These results indicate that the solid-state reaction in amorphous phase is rather close to reactions in solution and the molecular properties are maintained.

p-TsOH

trans-**13**   **14**   cis-**13**

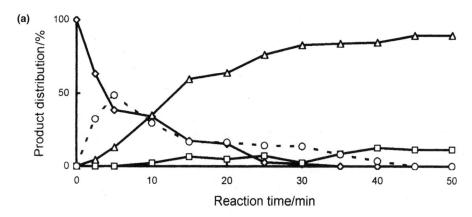

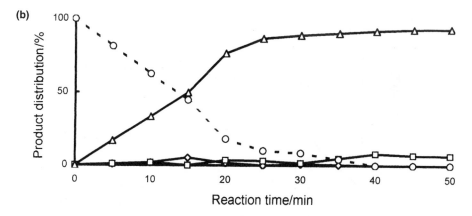

Figure 9 Time course of the reaction induced by grinding **13** with *p*-toluenesulfonic acid. (a) *trans*-**13**. (b) *cis*-**13**. —○— : *trans*-**13**   ----○---- : *cis*-**13**   —△— : **14** (from [21] (p. 81)).

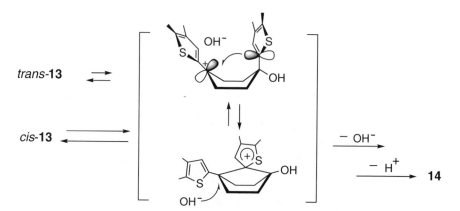

Figure 10 π-Participation in the 1,4-rearrangement and *cis-trans* equilibration of **13** (from [21] (p. 81)).

Another example of the reaction in the amorphous state is acid-catalyzed *cis-trans* isomerization of 9,10-di(3-thienyl)-9,10-dihydoxy-9,10-dihydroanthracene (**65**), in which no 1,4-pinacol rearrangement is induced. When a mixture of *trans*-**65** and *p*-toluenesulfonic acid in equimolar amounts is ground in a mortar with a pestle, solid-state isomerization to *cis*-**65** takes place in 70% yield [110]. Throughout the cogrinding no wetting or melting of the solids is observed, while the resulting solids are amorphous as revealed by powder X-ray diffraction. Similarly, cogrinding of *cis*-**65** gaves rise to the formation of *trans*-**65** in 28% yield (Figure 11). The equilibrium composition, 30:70 of *trans*:*cis*, is established whichever isomer of **65** is allowed to undergo the isomerization, indicating the existence of equilibration in the solid-state cogrinding. This composition is very close to that found for the equilibration in solution (80:20). In the non-crystalline

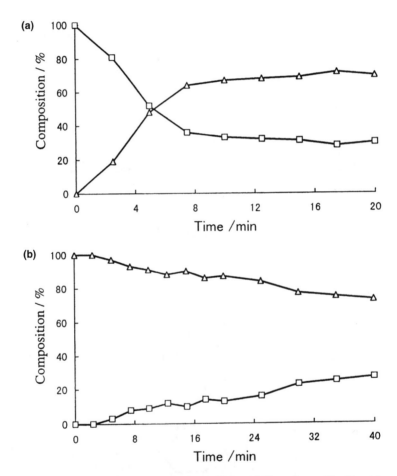

Figure 11 (a) Time course of the equilibration induced by cogrinding of *trans*-**65** with *p*-toluenesulfonic acid. □: *trans*-**65**. △: *cis*-**65**. (b) Time course of the equilibration induced by cogrinding of *cis*-**65** with *p*-toluenesulfonic acid. □: *trans*-**65**. △: *cis*-**65**.

state the intermolecular interactions and the molecular packing energies should be less important and, therefore, the molecules are able to maintain their molecular identities in the solid state.

trans-**65**   *cis*-**65**

11.   Oxidation and Reduction

The solid-state oxidations and reductions are carried out by cogrinding the substrate and the reagent. The reduction of ketones by NaBH$_4$ was achieved by grinding the mixture and keeping the resulting powder in a dry box at room temperature for 5 days, being stirred once a day [111]. Enantioselective reduction is realized in 100% ee by complex-ation of the ketones with chiral host compounds. The chiral host compounds **66** and **67** affords 1:1 including compounds with ketones as guest components. These inclusion crystals are mixed with 2BH$_3$-NH$_2$CH$_2$CH$_2$NH$_2$, finely powdered, and stirred. The corre-sponding *sec*-alcohols are enantioselectively obtained, although enantioselectivity is not very high [112].

(*R,R*)-(–)-**66**   (*R,R*)-(–)-**67**

Other enantioselective solid-state ionic reactions using chiral enclathration are reported by Toda and co-workers, which include Michael addition [113], Wittig–Horner reaction [114], and ylide reactions to give cyclopropanes and oxiranes [115].

Solid-state reductions of carbonyl groups in cage diketone **68** have been reduced stereospecifically by sodium boron hydride. Hydride transfer occurs exclusively at the exo face of the carbonyl group. In contrast, the corresponding reduction performed in ethanol displays only moderate steroselectivity [116]. Baeyer–Villiger oxidation of ketones with *m*-chloroperbenzoic acid proceeds much faster in the solid state than in solution [117].

$$\text{68} \xrightarrow{\text{NaBH}_4} \text{69}$$

68　　　　　　　　　69

12. Final Remarks

The experimental simplicity of solid-state reactions is a compelling reason to use these reactions as preparative procedures. The trouble or expense of providing and removing a solvent is avoided. Not only such practical usage but also the inherent chemistry associated with the solid-state ionic reaction is an interesting subject of research. Compared with the solid-state photoreactions, much still remains to be clarified in the mechanistic aspect of solid-state ionic reactions. The generation of the ions in the crystalline lattice, even though these are reaction intermediates, should affect the solid-state reactivity, since Coulombic interaction is stronger than van der Waals interaction and the positive and negative charges interact over long distance. The solid-state ionic reactions are also important from the viewpoint of material sciences. Physico-chemical properties of molecular solids such as charge-transfer, charge-carrier, dielectric properties, and ion-transport are all related to solid-state ionic reactions. The theoretical treatment of the solid-state ionic reactions has been rather extensively studied in inorganic compounds especially at high temperatures because of their industrial utility. Some results obtained in this field would be valid in the solid-state reactions of molecular crystals.

The discovery of new solid-state reactions are rather serendipitous. In order to design the solid-state reaction, crystal engineering is essential. In particular, for two-component reactions, it is necessary to use cocrystals containing two or more of the requisite compounds. Concerted efforts to design cocrystals will bring about fruitful results in two-component solid-state reactions. The reaction induced by gas-solid contact has been long known, and is related highly with the surface phenomenon. Modern supermicroscopic technology should be useful there. Another type of solid-state reaction is that induced by the cogrinding of two solids of the components. In this case, it is necessary to gain insight into the implication of grinding at the molecular level.

The unstable reaction intermediates, which are generated in crystalline lattices, can be persistent because of restricted molecular motion. This characteristic could be applied to a study on the generation of new labile ionic species and their reactivities in the solid state. Then, we have a good chance to discover new solid-state reactions.

References

1. V. RAMAMURTHY and K. VENKATESAN. *Chem. Rev.* **87** (1987) 433.
2. I. C. PAUL and D. Y. CURTIN. *Science* **187** (1975) 19.

3.   I. C. PAUL and D. Y. CURTIN. *Acc. Chem. Res.* **6** (1973) 217.
4.   F. TODA and K. TANAKA. *Chem. Rev.* **100** (2000) 1025.
5.   G. M. J. SCHMIDT. *Pure Appl. Chem.* **27** (1971) 643.
6.   M. D. COHEN and G. M. J. SCHMIDT. *J. Chem. Soc.* (1964) 1996.
7.   M. D. COHEN. *Angew. Che. Int. Ed. Engl.* **14** (1975) 386.
8.   T. LUTY and C. J. ECKHARDT. *J. Am. Chem. Soc.* **117** (1995) 2441.
9.   J. M. MCBRIDE. *Acc. Chem. Res.* **16** (1983) 304.
10. J. M. MCBRIDE, B. E. SEGMULLER, M. D. HOLINGSWORTH, D. E. MILLS and B. A. WEBER. *Science* **234** (1986) 830.
11. G. KAUPP. *Angew. Chem. Int. Ed. Engl.* **31** (1992) 592.
12. L. A. ERREDE, M. ETTER, R. C. WILLIAMS and S. M. DARNAUER. *J. Chem. Soc., Perkin Trans.* **2** (1980) 233.
13. T. W. LEWIS, D. Y. CURTIN and I. C. PAUL. *J. Am. Chem Soc.* **101** (1979) 5717.
14. N. B. SINGH, R. J. SINGH and N. P. SINGH. *Tetrahedron* **50** (1994) 6441.
15. T. KAWATO, H. KANATOMI, K. AMIMOTO, H. KOYAMA and H. SHIGEMIZU. *Chem. Lett.* (1999) 47.
16. K. TANAKA, T. WATANABE and F. TODA. *Chem. Commun.* (2000) 1363.
17. N. HAYASHI, Y. MAZAKI and K. KOBAYASHI. *Tetrahedron Lett.* **35** (1994) 5883.
18. M. TANAKA, S. HATADA, N. TANIFUJI and K. KOBAYASHI. *J. Org. Chem.* **66** (2001) 803.
19. M. TANAKA and K. KOBAYASHI. *Chem. Commun.* (1998) 1965.
20. M. DESSOLIN and M. GOLFIER. *J. Chem. Soc., Chem. Commun.* (1986) 38.
21. R. SEKIYA, K. KIYO-OKA, T. IMAKUBO and K. KOBAYASHI. *J. Am. Chem. Soc.* **122** (2000) 10282.
22. J. Z. GOUGOUTAS. *J. Am. Chem. Soc.* **101** (1979) 5697.
23. J. BOERIO-GOATES, J. I. ARTMAN and D. GOLD. *J. Phys. Chem. Solids* **48** (1987) 1185.
24. J. WALKER and J. K. WOOD. *J. Chem. Soc.* **77** (1900) 21.
25. R. E. PINCOCK, K. R. WILSON and T. E. KIOVSKY. *J. Am. Chem. Soc.* **89** (1967) 6890.
26. C. N. SUKENIK, J. A. P. BONAPCE, N. S. MANDEL, P-Y LAU, G. WOOD and R. G. BERGMAN. *J. Am. Chem. Soc.* **99** (1977) 851.
27. (a) J. A. R. P. SARMA and J. D. DUNITZ. *Acta Cryst.* **B46** (1990) 780. (b) A. GAVEZZOTTI. *J. Am. Chem. Soc.* **105** (1983) 5220.
28. M. ODA and N. SATO. *J. Phys. Chem. B* **102** (1998) 3283.
29. K. DWARAKANATH and P. N. PRASAD. *J. Am. Chem. Soc.* **102** (1980) 4254.
30. F. MENGER, H. B. KAISERMAN and L. J. SCOTCHIE. *Tetrahedron Lett.* **25** (1984) 2311.
31. J. BOERICO-GOATES, J. I. ARTMAN and D. GOLD. *J. Phys. Chem. Solids* **48** (1987) 1185.
32. J. A. R. P. SARMA and J. D. DUNITZ. *Acta Cryst.* **B46** (1990) 780 and 784.
33. M. ODA and N. SATO. *Chem. Phys. Lett.* **275** (1997) 40.
34. J. SCHMEYERS, F. TODA, J. BOY and G. KAUPP. *J. Chem. Soc., Perkin Trans.* **2** (1998) 989.
35. G. KAUPP and J. SCHMEYERS. *J. Org. Chem.* **60** (1995) 5459.
36. G. KAUPP and J. SCHMEYERS. *Angew. Chem. Int. Ed. Engl.* **32** (1993) 1587.
37. H. M. MCCONNELL, B. M. HOFFMAN and R. M. METZGER. *Proc. Natl. Acad. Sci. USA* **53** (1965) 46.
38. J. B. TRANCE, J. E. VAXQUEZ, J. J. MAYERLE and V. Y. LEE. *Phys. Rev. Lett.* **46** (1981) 253.
39. H. OKAMOTO, T. KODA, Y. TOKURA, T. MITANI and G. SAITO. *Phys. Rev. B* **39** (1989) 10693.
40. M. J. S. DEWAR and A. LEPLEY. *J. Am. Chem. Soc.* **83** (1961) 4560.
41. H. SATO and T. YASUHIWA. *Bull. Chem. Soc. Jpn.* **47** (1974) 368.
42. F. TODA and H. MIYAMOTO. *Chem. Lett.* (1995) 861.
43. V. R. PEDIREDDI, W. JONES, A. P. CHORLTON and R. DOCHERTY. *Chem. Commun.* (1996) 987.
44. D. BRAGA, L. MAINI and F. GREPIONI. *Chem. Commun.* (1999) 937.
45. J. W. EASTMAN, G. M. ANDROES and M. CALVIN. *J. Chem. Phys.* **36** (1962) 1197.
46. N. TANIFUJI and K. KOBAYASHI. unpublished results.
47. Y. MATSUNAGA. *Bull. Chem Soc. Jpn.* **49** (1976) 1411.
48. Y. ITO, H. NAKABAYASHI, S. OHBA and H. HOSOMI. *Tetrahedron* **56** (2000) 7139.
49. T. SUZUKI, T. FUKUSHIMA, Y. YAMASHITA and T. MIYAJI. *J. Am. Chem. Soc.* **116** (1994) 2793.
50. H. KOSHIMA, K. DING, Y. CHISAKA and T. MATSUURA. *J. Am. Chem. Soc.* **118** (1996) 12059.
51. H. KOSHIMA, D. MATSUSHIGE, M. MIYAUCHI and J. FUJITA. *Tetrahedron* **56** (2000) 6845.
52. N. HAGA, H. NAKAJIMA, H. TAKAYANAGI and K. TOKUMARU. *J. Org. Chem.* **63** (1998) 5372.
53. (a) R. S. MILLER, D. Y. CURTIN and I. C. PAUL. *J. Am. Chem. Soc.* **93** (1971) 2784.

(b) R. S. MILLER, D. Y. CURTIN and I. C. PAUL. *J. Am. Chem. Soc.* **94** (1972) 5117.

(c) R. S. MILLER, D. Y. CURTIN and I. C. PAUL. *J. Am. Chem. Soc.* **96** (1974) 6329, 6334, and 6340.

54. C-T. LIN, I. C. PAUL and D. Y. CURTIN. *J. Am. Chem. Soc.* **96** (1974) 3699.

55. E. N. DUESLER, R. B. KRESS, C-T. LIN, W-I. SHIAU, I. PAUL and D. Y. CURTIN. *J. Am. Chem. Soc.* **103** (1981) 875.

56. R. S. MILLER, D. Y. CURTIN and I. C. PAUL. *J. Am. Chem. Soc.* **93** (1971) 2784.

57. C-T. LIN, P-Y. SIEW and S. R. BYRN. *J. Chem. Soc., Perkin Trans.* **2** (1978) 963.

58. N. B. SINGH, N. P. SINGH, V. A. KUMAR and M. NETHAJI. *J. Chem. Soc., Perkin Trans.* **2** (1994) 361.

59. Y. TOYOSHI, T. NAKATO and T. OKUHARA. *Bull. Chem. Soc. Jpn.* **71** (1998) 2817.

60. F. TODA, H. TAKUMI and M. AKEHI. *J. Chem. Soc., Chem. Comm.* (1990) 1270.

61. G. KAUPP, M. HAAK and F. TPDA. *J. Phys. Org. Chem.* **8** (1995) 545.

62. N. ARIMOTO, N. TANIFUJI and K. KOBAYASHI. unpublished results.

63. G. KAUPP, U. POGODDA and J. SCHMEYERS. *Chem. Ber.* **127** (1994) 2249.

64. F. TODA, K. TANAKA, Y. KAGAWA and Y. SAKAINO. *Chem. Lett.* (1990) 373.

65. A. P. PATIL, D. Y. CURTIN and I. C. PAUL. *J. Am. Chem. Soc.* **106** (1984) 348.

66. A. P. PATIL, D. Y. CURTIN and I. C. PAUL. *J. Am. Chem. Soc.* **106** (1984) 4010.

67. K. NAKASUJI, K. SUGIURA, T. KITAGAWA, J. TOYODA, H. OKAMOTO, K. OKANIWA, T. MITANI, H. YAMAMOTO, I. MURATA, A. KAWAMOTO and J. TANAKA. *J. Am. Chem. Soc.* **113** (1991) 1862.

68. M. T. REETZ, S. HOGER and K. HARMS. *Angew. Chem. Int. Ed. Engl.* **33** (1994) 181.

69. M. TANAKA, H. MATSUI, J. MIZOGUCHI and S. KASHINO. *Bull. Chem. Soc. Jpn.* **67** (1994) 1572.

70. M. TANAKA and H. TSUNEKAWA. *Mol. Cryst. Liq. Cryst.* **313** (1998) 355.

71. J. R. DE LA VEGA. *Acc. Chem. Res.* **15** (1982) 185.

72. For tautomerism in the solid state, T. SUGAWARA and I. TAKASU. *Adv. Phys. Org. Chem.* (1999) 38.

73. (a) J. BREGMAN, L. LEISEROWITZ and G. M. SCHMIDT. *J. Chem. Soc.* (1964) 2068. (b) J. BREGMAN, L. LEISEROWITZ and K. OSAKI. *J. Chem. Soc.* (1964) 2086.

74. (a) T. INABE, N. HOSHINO, T. MIYANI and Y. MARUYAMA. *Bull. Chem. Soc. Jpn.* **62** (1989) 2245. (b) N. HOSHINO, T. INABE, T. MITANI and Y. MARUYAMA. *Bull. Chem. Soc. Jpn.* **61** (1988) 4207.

75. K. OGAWA, Y. KASAHARTA, Y. OHTANI and J. HARADA. *J. Am. Chem. Soc.* **120** (1998) 7107.

76. H. KOYAMA, T. KAWATO, H. KANATOMI, H. MATSUSHITA and K. YONETANI. *J. Chem. Soc., Chem. Commun.* (1994) 579.

77. T. KAWATO, H. KANATOMI, K. AMIMOTO, H. KOYAMA and H. SHIGEMIZU. *Chem. Lett.* (1999) 47.

78. Y. EICHEN, J-M. LEHN, M. SCHERL, D. HAARER, J. FISCHER, A. DECIAN, A. CORVAL and H. P. TROMMSDORFF. *Angew. Chem. Int. Ed. Engl.* **34** (1995) 2530.

79. B. FORCES-FORCES, A. L. LLAMAS-SAIZ, R. M. CLARAMUNT, C. LOPEZ and J. ELGUERO. *J. Chem. Soc., Chem. Commun.* (1994) 1143.

80. M. SMRCINA, S. VYSKOCIL, V. HANUS, M. POLASEK, V. LANGER, B. G. M. CHEW, D. B. ZAX, H. VERRIER, K. HARPER, T. A. CLAXTON and P. KOCOVSKY. *J. Am. Chem. Soc.* **118** (1996) 487.

81. F. TODA and K. OKUDA. *J. Chem. Soc., Chem. Commun.* (1991) 1212.

82. (a) K. KURUMA, H. NAKAGAWA, T. IMAKUBO and K. KOBAYASHI. *Bull. Chem. Soc. Jpn.* **72** (1999) 1395. (b) N. HAYASHI, K. KURUMA, Y. MAZAKI, T. IMAKUBO and K. KOBAYASHI. *J. Am. Chem. Soc.* **120** (1998) 3799. (c) N. HAYASHI, Y. MAZAKI and K. KOBAYASHI. *J. Org. Chem.* **60** (1995) 6432. (d) N. HAYASHI, Y. MAZAKI and K. KOBAYASHI. *J. Chem. Soc., Chem. Commun.* (1994) 2351.

83. D. S. GOUD and G. R. DESIRAJU. *J. Chem. Research (S)* (1995) 244.

84. J. A. R. SARMA, A. NAGARAJU, K. K. MAJUMDAR, P. M. SAMUEL, I. DAS, S. ROY and A. J. MCGHIE. *J. Chem. Soc., Perkin Trans.* **2** (2000) 1119.

85. M. C. ETTER, G. M. FRANKENBACH and J. BERNSTEIN. *Tetrahedron Lett.* **30** (1989) 3617.

86. R. W. TRIMMER, L. R. STOVER and A. C. SKJOLD. *J. Org. Chem.* **50** (1985) 3618.

87. J. Z. GOUGOUTAS and J. JOHNSON. *J. Am. Chem. Soc.* **100** (1978) 5816.

88. E. HADJOUDIS, E. KARIV and G. M. SCHMIDT. *J. Chem. Soc., Perkin Trans.* **2** (1972) 1056.

89. R. S. BROWN, R. W. NAGORSKI, A. J. BENNET, R. E. D. MCCLUNG, G. H. M. AARTS, M. KLOBUKOWSKI, R. MCDONALD and B. D. SANTARSIENO. *J. Am. Chem. Soc.* **116** (1994) 2448.

90. K. PENZIEN and G. M. J. SCHMIDT. *Angew. Chem. Int. Ed. Engl.* **8** (1969) 608.

91. (a) D. RABINOVICH and Z. SHAKKED. *Acta Cryst.* **B30** (1974) 2829.

(b) D. RABINOVICH and H. HOPE. *Acta Cryst.* **A36** (1980) 670.

92.   M. GARCIA-GARIBAY, J. R. SCHEFFER, J. TROTTER and F. WIREKO. *Tetrahedron Lett.* **29** (1988) 1485.

93.   Y. TANAKA, H. DAKURABA and H. NAKANISHI. *J. Chem. Soc., Chem. Commun.* (1983) 947.

94.   B. S. GOUD, K. PANNEERSELVAN, D. E. ZACHARIAS and G. R. DESIRAJU. *J. Chem. Soc., Perkin Trans.* **2** (1995) 325.

95.   (a) G-W. WANG, K. KOMATSU, Y. MURATA and M. SHIRO. *Nature* **387** (1997) 583. (b) K. KOMATSU, G-W. WANG, Y. MURATA, T. TANAKA, K. FUJIWARA, K. YAMAMOTO and M. SAUNDERS. *J. Org. Chem.* **63** (1998) 9358.

96.   K. KOMATSU, K. FUJIWARA and Y. MURATA. *Chem. Commun.* (2000) 1583.

97.   T. W. LEWIS, E. N. DUESLER, R. B. KRESS, D. Y. CURTIN and I. C. PAUL. *J. Am. Chem. Soc.* **102** (1980) 4659.

98.   S. A. PUCKETT, I. C. PAUL and D. Y. CURTIN. *J. Am. Chem. Soc.* **98** (1976) 787.

99.   G. FRIEDMAN, M. LAHAV and G. M. J. SCHMIDT. *J. Chem. Soc., Perkin Trans.* **2** (1974) 428.

100.   J. D. DUNTZ, K. D. HARRIS, R. L. JOHNSTON, B. M. KRIUKI, E. J. MACLEAN, K. PSALLIDAS, W. B. SCHWEITZER and R. R. TYKWINSKI. *J. Am. Chem. Soc.* **120** (1998) 13274.

101.   M. DESSOLIN and M. GOLFIER. *J. Chem. Soc., Chem. Commun.* (1986) 38.

102.   M. DESSOLIN, O. EISENSTEIN, M. GOLFIER, T. PRANGE and P. SAUTET. *J. Chem. Soc., Chem. Commun.* (1992) 132.

103.   M. L. TOSATO and L. SOCCORSI. *J. Chem. Soc., Perkin Trans.* **2** (1982) 1321 and 1593.

104.   M. KAFTRY and E. H-BENORY. *Mol. Cryst. Liq. Cryst.* **240** (1994) 241.

105.   E. H-BENORY, M. BOTOSHANSKY, M. GREENBERG, V. SHTEINMAN and M. KAFTORY. *Tetrahedron* **56** (2000) 6887.

106.   M. VICTORIA, M-DIAZ, R-MORGADE, W. SCHAFER and T. TORRES. *Tetrahedron* **49** (1993) 2261.

107.   K. VYAS and H. MANOHAR. *Mol. Cryst. Liq. Cryst.* **137** (1986) 37.

108.   K. VYAS, H. MAMANOHAR and K. VENKATESAN. *J. Phys. Chem.* **94** (1990) 6069.

109.   F. TODA and T. SHIGEMASA. *J. Chem. Soc., Perkin Trans.* **1** (1989) 209.

110.   R. SEKIYA and K. KOBAYAHSI. unpublished results.

111.   F. TODA, K. KIYOSHIGE and M. YAGI. *Angew. Chem. Int. Ed. Engl.* **28** (1989) 320.

112.   F. TODA and K. MORI. *J. Chem. Soc., Chem. Commun.* (1989) 1245.

113.   F. TODA, K. TANAKA and J. SATO. *Tetrahedron: Asymmetry* **4** (1993) 1771.

114.   F. TODA and H. ARAI. *J. Org. Chem.* **55** (1990) 3446.

115.   F. TODA and N. IMAI. *J. Chem. Soc., Perkin Trans.* **1** (1994) 2673.

116.   A. MARCHAND and G. M. REDDY. *Tetrahedron* **47** (1991) 6571.

117.   F. TODA, M. YAGI and K. KIYOSHIGE. *J. Chem. Soc., Chem. Commun.* (1988) 958.

Organic Photoreaction in the Solid State

Koichi Tanaka and Fumio Toda

Department of Applied Chemistry, Faculty of Engineering, Ehime University, Matsuyam, Ehime 790-8577, Japan

1. Introduction

Since molecules in crystals are arranged regularly, selective photoreactions occur efficiently upon photoirradiation in the solid state. When enantioselective reactions are required, achiral molecules should be arranged in a chiral form by using the chiral host molecules. Photoreaction of the chiral inclusion crystal gives an optically active product. In some cases, the enantioselective photoreaction proceeds in a single-crystal-to-single-crystal manner without collapse of the crystalline lattice during the photoreaction. Some achiral molecules are arranged in a chiral form in their own crystals and afford optically active products upon photoirradiation in the solid state.

   This chapter describes the stereoselective and enantioselective photoreactions of achiral molecules in inclusion crystals with various kinds of host compounds (**1–10**) and enantioselective photoreactions in their own chiral crystals.

F. Toda (ed.), Organic Solid-State Reactions, 109–158.
© 2002 *Kluwer Academic Publishers. Printed in Great Britain.*

OH

Ph

(C₆H₁₁)₂NOC

6

CON(C₆H₁₁)₂

7

Ph Ph
Me O OH
Me O OH
Ph Ph

8

Ph Ph
O OH
O OH
Ph Ph

9

Ph Ph
O OH
O OH
Ph Ph

10

2. Stereoselective [2+2] and [4+4] Photodimerization Reactions in Host-Guest Inclusion Crystals

The photodimerization of chalcone **11** is not easy in either solution or the solid state. For example, neither form I (mp 59 °C) [1] nor form II (mp 56 °C) [2] of chalcone **11** gives any dimer by photoirradiation in the solid state, although irradiation of **11** in EtOH for 147 h gives **14a**, albeit in only 9% yield [3].

Photodimerization of chalcone **11** was found, however, to undergo stereoselective control in inclusion crystals with host compound **1** [4]. For example, irradiation of a 1:2 inclusion complex of chalcone **11** with 1,1,6,6-tetraphenylhexa-2,4-diyne-1,6-diol **1**, in the solid state for 6 h, gave only the syn-head-to-tail dimer **12a** (of the four possible dimers **12–15**) in 82% yield. X-ray crystal structure analysis of the complex disclosed that two chalcone molecules **11** are packed close together through hydrogen bonds between the hydroxyl group of **1** and carbonyl oxygen of **11** in the complex (Figure 1). The double bonds are parallel and the distance between the two is 3.862 Å (Figure 2). It was also remarkable that photodimerization of chalcone proceeds stereoselectively in the molten state and gives the corresponding rac-anti-head-to-head dimers **14a** in relatively good yields [5]. Irradiation of one polymorphic crystal of chalcone **11** (form II, mp 56 °C) at 60 °C for 24 h by a 400-W high-pressure Hg lamp gave the anti-head-to-head dimer **14a** in 31% yield. Similar irradiation of another polymorphic crystal form of chalcone **11** (form I, mp 59 °C) at 60 °C also gave **14a** in 28% yield. The molecular structures of **12a** and **14a** were determined by X-ray analysis (Figure 3).

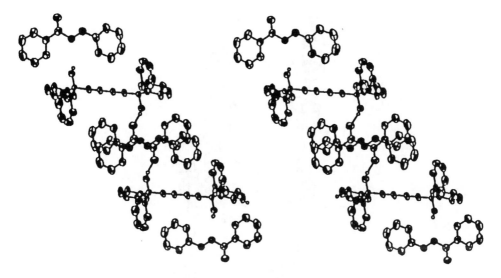

Figure 1 Stereoview of the 1:2 inclusion complex of **1** with **11**. Reprinted with permission from *J. Org. Chem.* **50** (1985) 2155. © American Chemical Society.

Table 1 Yields of the photocycloaddition reaction products

Host	Guest	Reaction time (h)	Product	Yield (%)
1	11	6	12a	90
6	11	6	12a	85
6	16	6	12b	70
1	17	8	18	86

The hydroquinone host compound **6** is also effective in assisting photodimerization of **11** or dibenzylideneacetone **16**. For example, irradiation of the 1:2 inclusion complex of **16** with **6** in the solid state for 6 h gave only the syn-head-to-tail dimer **12b** in 70% yield (Table 1). X-ray crystal analysis shows that hydrogen bonding plays an important role in packing the guest molecules **16** with the crystal, and the resulting distance between the double bonds of **16** is short enough (3.787 Å) to allow easy reaction [4].

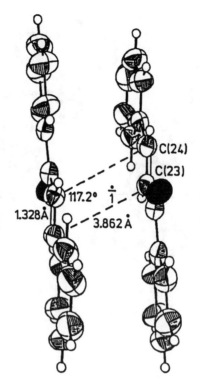

Figure 2   Mutual relation and geometrical parameters of the reacting centers of the guest molecules **11**. Reprinted with permission from *J. Org. Chem.* **50** (1985) 2155. © American Chemical Society.

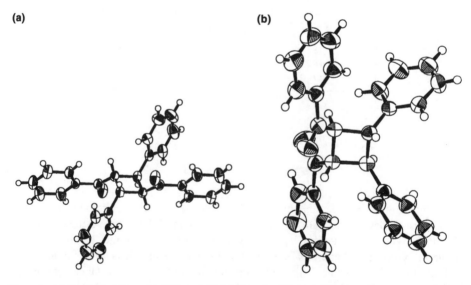

Figure 3   Molecular structures of (a) **12a** and (b) **14a**. Reprinted from *JCS Perkin Trans 1* (1998) 1316, by permission of the Royal Society of Chemistry.

Irradiation of the 1:2 inclusion complex of 9-formylanthracene **17** with **1** for 8 h in the solid state afforded the anti-dimer **18** in 86% yield. Two molecules of **17** are arranged between two host molecules through hydrogen bonding in an orientation that leads to the anti-dimer **18** on photodimerization and where the distance between the two reaction centers is short enough (4.042 Å) to permit this [4].

Since 2-pyridone **19a** exists as an equilibrium mixture with 2-hydroxypyridine **20a**, it is difficult to isolate **19a** in a pure state. However, inclusion complexation can be applied to the isolation of the keto-form in a pure form. For example, the host compound **1** selectively includes the keto-form **19a** to form a 1:2 inclusion crystal. This selective inclusion of **19a** can be used for its efficient dimerization. Irradiation of the 1:2 inclusion complex composed of **1** and **19a** in the solid state for 6 h gave the trans-anti-dimer

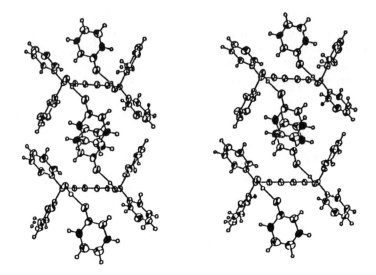

Figure 4 Stereoview of the 1:2 inclusion complex of **1** with **19a**. Reprinted from *Tetrahedron* **43** (1987) 1507, with permission from Elsevier Science.

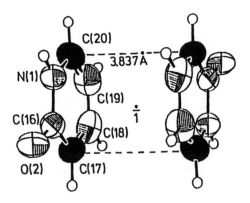

Figure 5 Mutual relation and geometrical parameters of the reactivity centers of the guest molecules **19a**. Reprinted from *Tetrahedron* **43** (1987) 1507, with permission from Elsevier Science.

22a in 76% yield [6]. In contrast to this, solution photodimerization of **19a** in EtOH gave **22a** only in 18% yield after longer (100 h) irradiation. X-ray crystal structural study of the complex shows that two molecules of **19a** are arranged between the host molecules in positions that allow formation of the trans-anti-dimer **22a** by dimerization (Figure 4) since the distance between the reaction centers is very close (3.837 Å) (Figure 5). Similar photodimerizations of N-methylpyridones **21b**, **21c**, **21e**, **21g**, **21h**, and **21j** also proceed efficiently in inclusion crystals rather than in solution, although the inclusion complexes **19c–j** are photostable (Table 2).

a: R^1 = R2 = R3 = R4 = H
b: R2 = R3 = R4 = H, R1 = Me
c: R1 = R3 = R4 = H, R2 = Me
d: R1 = R2 = R4 = H, R3 = Me
e: R1 = R2 = R3 = H, R4 = Me
f: R2 = R4 = H, R1 = R3 = Me

g: R1 = R3 = H, R2 = R4 = Me
h: R1 = R2 = H, R3 = R4 = Me
i: R1 = H, R2 = R3 = R4 = Me
j: R1 = R2 = R3 = H, R4 = OMe

19   **20**   **21**

22   **23**   **24**

Table 2 Yields of photoreaction products of 2-pyridones **19** and 1-methyl-2-pyridones **21** in the crystalline complex and in EtOH solution

Substituent	Yield (%)[a]			
	22		**23**	
	Complex	EtOH soln	Complex	EtOH soln
a	76	18	32	30
b	(22)[b]	18	19	30
c	1	7	85	21
d	0	45	0	41
e	0	9	74	9
f	0	69	1	40
g	0	10	74	3
h	0	40	91	37
i	0	32	9	19
j	0	0	84	0

[a] The irradiation times were 6 h for Complexes and 100 h for ethanol Solution.
[b] The intermolecular dimer **24** was obtained.

3. Enantioselective Photoreactions in Host-Guest Inclusion Crystals

It has been found that enantioselective photoreactions can be successfully controlled in crystalline inclusion complexes by using optically active host compounds. For example, disrotatory intramolecular [2+2] photocyclization of tropolone alkyl ethers can be controlled perfectly by using optically active hosts. Irradiation of a 1:1 complex of α-tropolone methyl ether **25a** and (S,S)-(−)-**3** in the solid state gave (1S,5R)-(−)-**26a** of 100% ee and (+)-**28a** of 91% ee in 11 and 26% yields, respectively [7]. Similar irradiation of a 1:1 complex of **25b** with (−)-**3** gave (1S,5R)-(−)-**26b** of 100% ee and (+)-**28b** of 72% ee in 12 and 14% yields, respectively. The enantioselective photoreaction can be interpreted as follows: disrotatory [2+2] photoreaction of **25** in the inclusion crystals

115

K. Tanaka and F. Toda

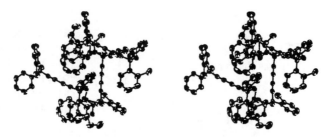

Figure 6   Stereoscopic view of the 1:1 inclusion complex of (–)-3 and 25 showing the hydrogen bond network. Reprinted with permission from *J. Org. Chem.* **53** (1988) 4392. © American Chemical Society.

with (–)-3 occurs only in the A direction due to the steric hindrance of (–)-3. This interpretation was deemed to be reasonable through an X-ray crystal structure study of the inclusion complex (Figure 6) [8].

(S,S)-(–)-   Ph⊢ ≡ ≡ ⊣Ph   ·

3

B 25

1:1 complex

A   / hν \   B

a: R = Me; b: R = Et

$(1S,5R)$-(–)-**26**

$(1R,5S)$-(+)-**27**

—CH₂COOR

Similar disrotatory [2+2] photocyclization of pyridones into β-lactam derivatives also proceeds efficiently within inclusion crystals [9]. Irradiation of either powdered inclusion complexes of N-methylpyridone **29** with (–)-**8** or N-methyl-4-methoxypyridone **30** with (–)-**3** gave enantiomerically pure (+)-**31** or (–)-**32** in 49 and 8% yields, respectively.

Optically active oxazolidinones **34a–d** of 9.5–100% ee were obtained by irradiating the 1:1 inclusion complexes of nitrones **33a–d** and optically active host compound (–)-**13** in the solid state (Table 3) [10].

It has been reported that irradiation of N,N-diisopropylpyruvamide **35a** in benzene gives the oxaziridine **37a** exclusively and that irradiation in the solid state at $-78\,^{\circ}$C gives β-lactam **36a** and **37a** in 31 and 29% yields, respectively [11]. On the other hand, irradiation of the 1:1 inclusion compound of **35a** and **7** in the solid state for 40 h gave only **36a** in 60% yield. Similar irradiation of the 1:1 inclusion compound of **35b** and **7** gave **36b** selectively in 56% yield [12].

Table 3 Yield and optical purity (% ee) of oxaziridines **34a–d**

	Ar	Yield (%)	Optical Purity (% ee)
a	Ph	41	9.5
b		74	30
c		51	100
d		52	94

35 a; $R^1 = Pr^i$
b; $R^1 = Me$

36 a; $R^1 = Pr^i$; $R^2 = Me$
b; $R^1 = Me$; $R^2 = H$

37

Stereoselective and enantioselective photoreactions of the N,N-dialkylphenylglyoxy-lamides **38** were accomplished by irradiation of their respective inclusion complexes with optically active host compounds **3**, **5**, **9** or **10** (Table 4) [13]. For example, when a 1:1 inclusion complex of (–)-**3** with **38a** was irradiated by a 400-W high-pressure Hg lamp at room temperature for 24 h, optically pure β-lactam (–)-**39a** was obtained in 90% yield. X-ray crystal structural studies showed that the conformation of **38a** is fixed in the solid state by the formation of two hydrogen bonds so as to give (–)-**39a**, as shown in Figure 7. Enantiocontrol of the reaction was also achieved in inclusion crystals using the optically active host compounds **5**, **9** and **10** [14]. For example, irradiation of the powdered 2:1 complex of **38a** with (–)-**10** gave (–)-**39a** of 100% ee in 40% yield. Similar photoirradiation of the 2:1 inclusion complex of **38d** with (–)-**10** gave (–)-**39d** of 100% ee in 11% yield and (–)-**40d** of 39% ee in 20% yield. However, photoreaction of **38h** as an inclusion complex with either (–)-**9** or (–)-**10** proceeded very selectively and gave only **40h** of 100% ee in quantitative yield (Table 4).

38

a: $R^1=R^2=R^3=R^4=H$
b: $R^1=R^3=H$; $R^2=R^4=Me$
c: $R^1=R^3=H$; $R^2=R^4=Et$
d: $R^1=R^2=H$; $R^3=R^4=Me$

e: $R^1=H$; $R^2=Me$; $R^3=R^4=Me$
f: $R^1=R^2=R^3=R^4=Me$
g: $R^1=R^3=H$; $R^2, R^4=-(CH_2)_3-$
h: $R^1=R^3=H$; $R^2, R^4=-CH_2OCH_2-$

39

40

Photocyclization reactions of the N-(aryloylmethyl)-δ-valerolactams **41** occur stereoselectively and enantioselectively in their inclusion crystals with optically active hosts [15]. For example, irradiation of the powdered inclusion complex of **41a** and the optically active host compound (–)-**10**, as a suspension in water for 12 h, gave (+)-**42a** of 98% ee in 59% yield. By similar irradiations of the 1:1 inclusion compounds of **41b** and **41c** with (–)-**10**, (–)-**42b** of 84% ee and (–)-**42c** of 98% ee were also obtained enantioselectively (Table 5). Conversely, photoreaction of **41a–c** in t-BuOH afforded a mixture of rac-**42a–c** and rac-**43a–c**. To elucidate the mechanism of the enantioselective photocyclization of **41**, the X-ray structure of the 1:1 complex of **41c** with **10** was determined (Figure 8) [16]. The molecules of **41c** are arranged in a chiral form within the crystal so as to give (6S,7S)-(–)-**42c** selectively. The distances for C1$\cdots$C3 and O1$\cdots$H3B are 3.24 and 2.82 Å, respectively, and hence the H3b atom is abstracted by O1 to form a hydroxyl group. The C3 atom attacks C1 from the re-face of the carbonyl group to give (6S,7S)-(–)-**42c**.

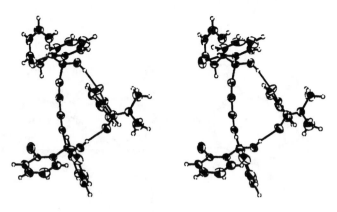

Figure 7 Stereoscopic view of the 1:1 inclusion complex of (–)-**3** and **38a**. Reprinted with permission from *J. Org. Chem.* **53** (1988) 4392. © American Chemical Society.

Table 4 Photoreaction of the phenylglyoxylamides **38** as inclusion complexes with the hosts **3, 9 and 10**

Host	Amide	39			40		
			Yield (%)	% ee		Yield (%)	% ee
(–)-**3**	**38a**	(–)-**39a**	90	100			
(–)-**3**	**38g**	(–)-*cis*-**39g**	29	63			
		(–)-*trans*-**39g**	21	95			
(–)-**3**	**38h**	(–)-*cis*-**39h**	31	56			
		(–)-*trans*-**39h**	9	a			
(–)-**9**	**38a**	(–)-**39a**	40	67	(–)-**40a**	55	100
(–)-**10**	**38a**	(–)-**39a**	40	100			
(–)-**9**	**38d**	(–)-**39d**	11	100	(–)-**40d**	20	39
(–)-**10**	**38d**	(+)-**39d**	17	61	(–)-**40d**	71	43
(–)-**9**	**38g**	(+)-*cis*-**39g**	17	96	(–)-**40g**	28	95
		(+)-*trans*-**39g**	32	44			
(–)-**10**	**38g**	(+)-*cis*-**39h**	15	100	(–)-**40g**	9	52
		(+)-*trans*-**39h**	34	95			
(–)-**9**	**38h**				(–)-**40h**	100	100
(–)-**10**	**38h**				(–)-**40h**	100	100
(–)-**5**	**38a**	(–)-**39a**	23	94			

ª % ee was not det determined.

41
a: X=H
b: X=Cl
c: X=Br

The chiral arrangement of **41a** molecules in (–)-**10** is also easily detected by measurement of the CD spectra as Nujol mulls. For example, the inclusion crystal of **41a** with

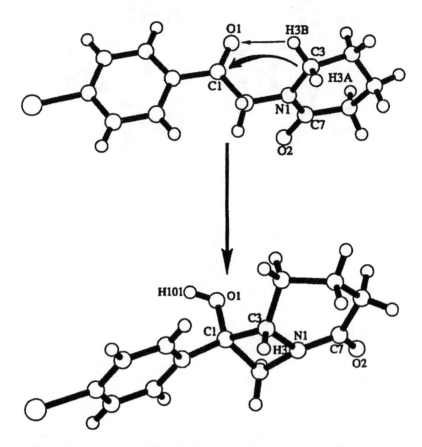

Figure 8 Photocyclization process of **41c** to **42c** in the inclusion crystals with (−)-**10**. Reprinted from *Bull. Chem. Soc. Jpn.* (1994) 2086, with permission from the Chemical Society of Japan.

Table 5 Photoreaction of **41** in the inclusion crystal and in solution

41	Host	Irradiation time (h)	42			43	
			Yield (%)		% ee	Yield (%)	% ee
41a	(−)-**10**	12	(+)-**42a**	59	98		
41a	(+)-**10**	12	(−)-**42a**	54	99		
41a	——— a	27	(±)-**42a**	33	0	(±)-**43a** 50	0
41b	(−)-**10**	12	(−)-**42b**	45	84		
41b	——— a	24	(±)-**42b**	34	0	(±)-**43b** 50	0
41c	(−)-**10**	15	(−)-**42c**	42	98		
41c	——— a	27	(±)-**42c**	26	0	(±)-**43c** 45	0

[a] Irradiation was carried out in *t*-BuOH.

(−)-**10** showed a (−)-Cotton effect, and that of **41a** with (−)-**10** showed a (+)-Cotton effect at around 260 and 310 nm, respectively, although the host molecule itself does not show any CD absorption (Figure 9).

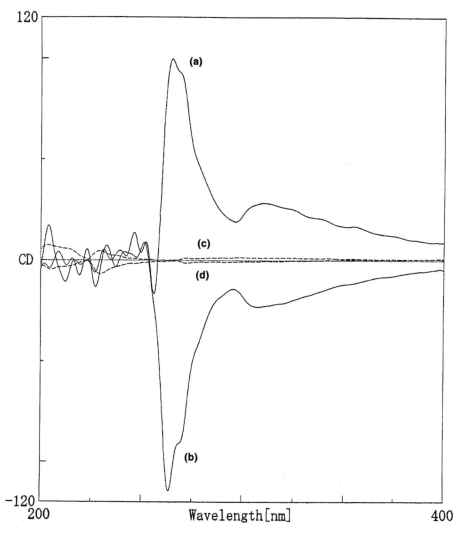

Figure 9 CD spectra of 1:1 inclusion complex of (a) **41a** with (+)-**10** and (b) **41a** with (–)-**10**, and of host molecule (c) (+)-**10** and (d) (–)-**10** in Nujol mulls. Reprinted with permission from *Chem. Rev.* (2000) 1059. © American Chemical Society.

The photocyclization reaction of acrylanilide to 3,4-dihydroquinoline was first reported in 1971 and its application to alkaloid synthesis has long been studied [17]. Although stereocontrol and enantiocontrol are important in this reaction, no such attempt has been reported, except one enantioselective photocyclization of 1-methylacryl-N-methylanilide **44a** in benzene-ether containing (+)-di(p-toluoyl)tartaric acid to give (–)-3-methyl-N-methyl-3,4-dihydroquinoline **46a** of 12–16% ee [18]. Recently, the host-guest inclusion method was found to be useful for a selective photocyclization of acrylanilides. For example, irradiation of the 1:1 inclusion compound of **44a** with (–)-**9** gave (–)-**46a** of 98% ee in 46% yield. On the other hand, similar irradiation of the 1:1 inclusion compound of **44a** with (−)-**10** gave (+)-**46a** of 95% ee in 29% yield (Table 6) [19]. It is surprising that the two host compounds (−)-**9** and (−)-**10** of so little structural difference (i.e. five-membered ring and six-membered ring) caused such disparate enantioselectivity. The 1,5-hydrogen shift in the photocyclizations of **44b–c** in its inclusion crystals with (−)-**9** or (−)-**10** is also controlled precisely and finally gives the trans-isomer **46b–c** of high optical purity (Table 6). When the photoirradiations of **44b–c** were carried out in solution, a 1:1 mixture of rac-**46b–c** and rac-**47b–c** was obtained.

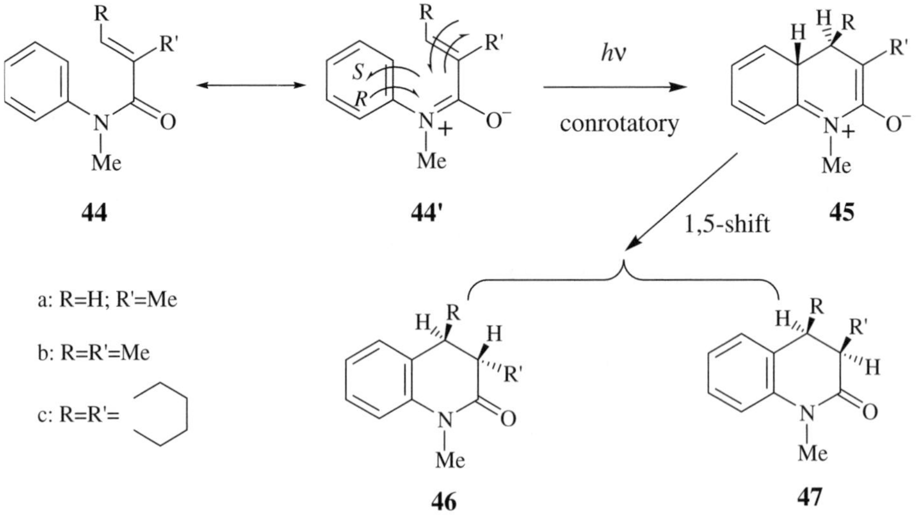

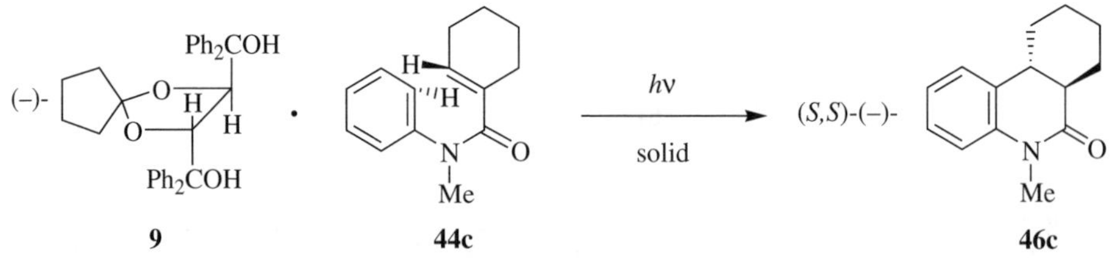

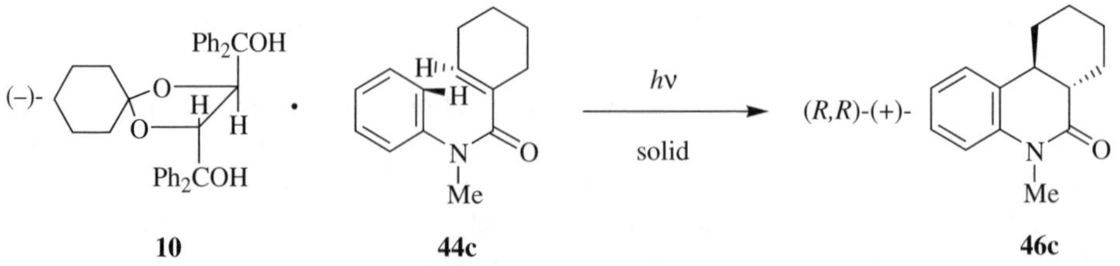

a: R=H; R'=Me

b: R=R'=Me

c: R=R'=

The selective conversion of **44** to **46** in the inclusion crystal can be interpreted as follows: of the two possible directions (*S* and *R*) of the conrotatory ring closure of the iminium form (**44'**) of **44**, only the rotation toward *S*, for example, occurs through control

Table 6 Photocyclization of anilides in the inclusion crystals

Anilides	Host	Yield (%)		% ee
44a	(−)-**9**	(−)-**46a**	46	98
44a	(−)-**10**	(+)-**46a**	29	95
44b	(−)-**9**	(−)-**46b**	65	98
44b	(−)-**10**	(+)-**46b**	44	98
44c	(−)-**9**	(+)-**46c**	62	70
44c	(−)-**10**	(−)-**46c**	29	99

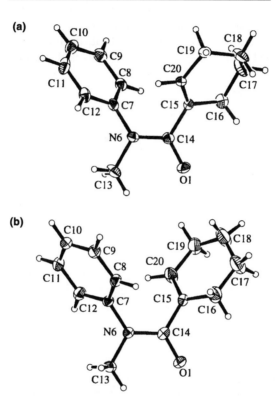

Figure 10 The molecular structures of guest **44c** in the 1:1 inclusion complex with (a) (−)-**9** and (b) (−)-**10**. Reprinted from *Bull. Chem. Soc. Jpn.* (2000) 2076, with permission from the Chemical Society of Japan.

by the host **9** (or **10**) to give the intermediate **45**. 1,5-hydrogen shift of the intermediate **45** occurs in a suprafacial manner to give the optically active photocyclization product **46**. The chiral conformation of **44** in the inclusion crystals with **9** and **10** was determined by X-ray analysis [20]. For example, the molecular structure of **44c** in its inclusion crystals with (−)-**9** is enantiomorphic to that in the inclusion crystals with (−)-**10**, as shown in Figure 10. In the inclusion crystal of **44c** with (−)-**9**, conrotatory photocyclization of **44c** with a positive torsion angle for C7-N6-C14-C15 will occur in a way that the hydrogens at C8 and C20 do not collide. Afterwards, a 1,5-hydrogen shift will occur in a suprafacial

123

manner to give (R,R)-(+)-**46c**. (S,S)-(–)-**46c** will be obtained in a similar way upon photo-irradiation of **44c** with a negative torsion angle for C7-N6-C14-C15 in its inclusion crystal with (–)-**10**. These assignments are supported by the known absolute configuration of the photoproducts [17].

Contrarily, the conformation of **44b** in the inclusion crystals with (–)-**9** is not suitable for cyclization, as shown in Figures 11a and 12a, although the enantioselective photo-cyclization occurred very efficently. Thus it is apparent that the intramolecular rotation of the methylmethacryl moiety of anilide **44b** actually occurred inside the crystal before cyclization [21]. When the crystals of the 1:1 inclusion complex of **44b** and (–)-**9** were irradiated with a 250-W ultrahigh-pressure Hg lamp through band-pass filter BP365 for 143 h, a disordered structure of the guest in the inclusion crystals was obtained, as shown in Figure 11b and 12b. The solid lines indicate the original structure of **44b**. The open lines correspond to the cyclized product, (–)-**46b**, with occupancy 26(1)%.

The reaction cavity of guest **44b** in the inclusion crystal is shown in Figure 13a. The positions of the C16 and C18 atoms will be exchanged by intramolecular rotation, and the C17 atom will be accommodated in the cavity. Figure 13b is a projection of the cavity around the C16, C17, and C18 atoms perpendicular to the C14-C15 rotation axis. By photoirradiation, the temperature of the crystal will be increased, and the space becomes sufficient for 180° rotation. Photocyclization of a guest molecule will alter the reaction cavity of the neighboring unit cell, and the photoreaction may proceed in domino fashion.

Photoirradiation of inclusion crystals of 3-oxo-2-cyclohexanecarboxamide **48a** with the optically active host compound **9** as a water suspension for 4 h gave optically active 2-aza-3,3-dimethyl-1,5-dioxaspiro[3,5]nonane **49a**. Optically pure **49b** and **49c** can be prepared by a combination of photoirradiation of **48b** and **48c**, respectively, in their inclusion crystals with **9** and purification of the reaction product via the oxime derivatives **50b** and **50c** [22].

(a)

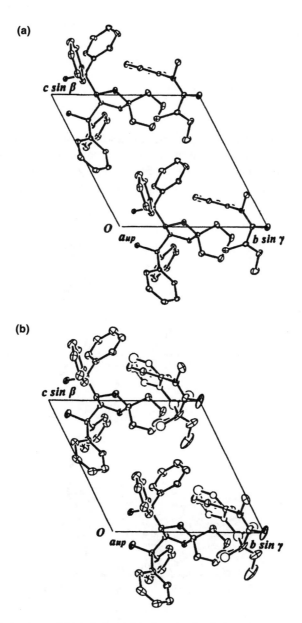

(b)

Figure 11 Crystal structures of (a) before and (b) after photoirradiation of the 1:1 inclusion complex of **44c** and (–)-**9**. Reprinted with permission from *J. Am. Chem. Soc.* **122** (2000) 1819. © American Chemical Society.

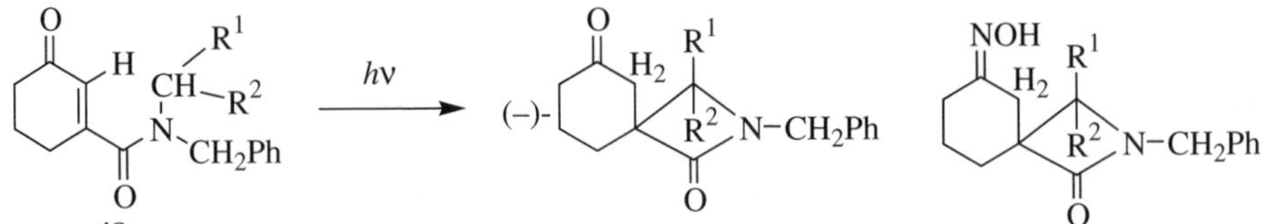

48

a: $R^1=R^2=H$ c: $R^1=H$; $R^2=Et$
b: $R^1=H$; $R^2=Me$ d: $R^1=R^2=Me$

2-(N-benzoyl-N-benzylamino)cyclohex-2-enones **51a–c** form inclusion crystals with optically active host compounds **10** in 1:1 or 2:1 ratios. Photoirradiation for 8 h with stirring of a 2:1 inclusion crystal of **10** and **51a** in water gave (–)-**52a** in 97% ee in 42% yield. Similarly, **51b** and **51c** are converted to the optically active β-lactams **52b** and **52c** in high optical purities [23].

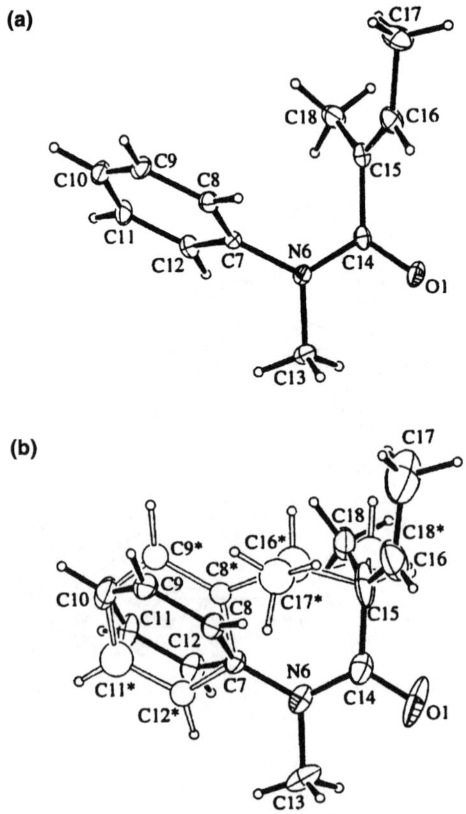

Figure 12 Molecular structure of **44c** in (a) before and (b) after photoirradiation of the 1:1 inclusion complex of **44c** and (–)-**9**. Atoms with asterisks in (b) have a site occupancy factor of 26(1)%. Reprinted with permission from *J. Am. Chem. Soc.* **122** (2000) 1818. © American Chemical Society.

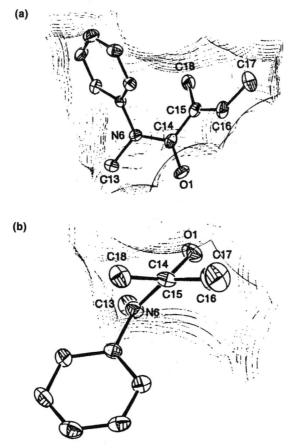

51 a: R_1 = PhCH$_2$, R_2 = Ph, R_3 = H   **52**
     b: R_1 = PhCH$_2$, R_2 = Me, R_3 = H
     c: R_1 = i-Pr, R_2=R_3= -(CH$_2$)$_5$-

Intramolecular [2+2] photocyclization reactions are also controlled enantioselectively within inclusion crystals. Photoirradiation of inclusion crystals of the 2-[N-(2-propenyl)-amino]cyclohex-2-enones **53** with the optically active host compounds **9** or **10** in the solid state gave the optically active 9-azatricyclo[5.2.1.0 [1,6]]decan-2-ones **54** (Table 7) [22].

Figure 13   (a) Sections of the reaction cavity of guest **44c** parallel to the plane defined by O1, C14, and C15, with a shift from −1 to +1 Å. (b) Sections of the reaction cavity perpendicular to the C14–C15 bond axis and through the C16 or the C17 atom. Positions of C16 and C18 will be exchanged by the intramolecular rotation of the methylmethacryl group. Reprinted with permission from *J. Am. Chem. Soc.* **122** (2000) 1819. © American Chemical Society.

Table 7 Photocyclization of **53** in a 2:1 inclusion compound with **9** and **10**

Host	Guest	Product	Yield (%)	$[\alpha]_D$ (deg) (c in CCl$_4$)
(−)-**9**	**53a**	**54a**	64	+70 (0.10)
(−)-**10**	**53a**	**54a**	81	+90 (1.32)
(−)-**10**	**53b**	**54b**	71	+77 (0.85)
(−)-**10**	**53c**	**54c**	65	+61 (0.53)

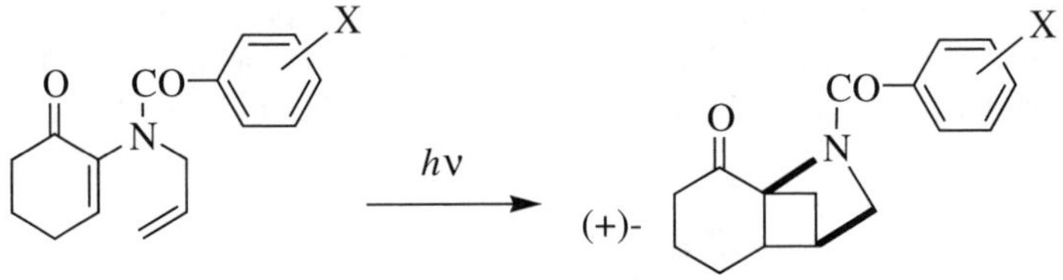

53 a: X=H
b: m-Cl
c: p-Me

54

Photoirradiation of inclusion crystals of the 4-(3-butenyl)cyclohexa-2,5-dien-1-ones **55** with the optically active host compounds **10** in the solid state gave optically active 1-carbomethoxytricyclo-[4.3.1.0 [7,10]]dec-2-en-4-ones **56** [22]. For example, reaction of a powdered 2:1 inclusion crystal of **55a** with (−)-**10** as a water suspension for 5 h gave (+)-**56a** of 73% ee in 50% yield. In the case of **55b**, enantioselective inclusion complexation occurred to give a 1:1 complex of optically pure (−)-**55b** with (−)-**9**, and its irradiation in a water suspension gave optically pure (+)-**56b** in 57% yield.

Photoirradiation of the 1:2 inclusion compounds of **57a–g** with (−)-**10** gave the optically active photocycloaddition products **58a–g** in high optical purity. However, photoirradiation of the 1:2 inclusion compound of **57c** and (−)-**10** gave the optically active spiro β-lactam **59c** of 97% ee in 69% yield (Table 8) [24].

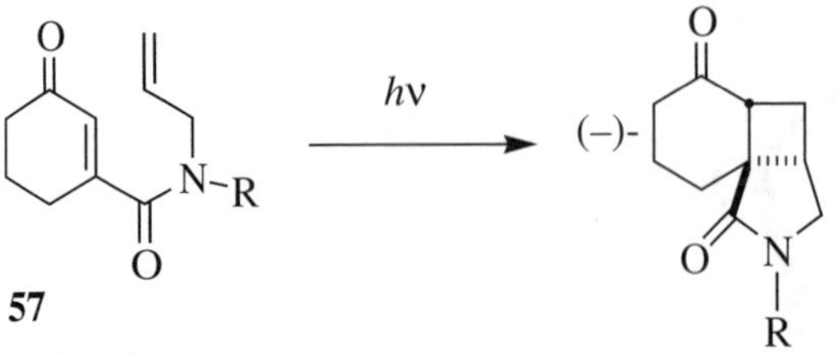

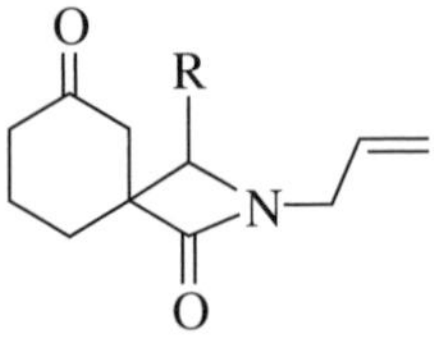

57
a: R=Me e: R=C$_6$H$_5$CH$_2$
b: R=Et f: R=p-MeC$_6$H$_4$CH$_2$
c: R=Pr^n g: R=p-ClC$_6$H$_4$CH$_2$
d: R=Bu^n

58

59

Table 8   Photocyclization of **57** in a 1:2 inclusion compound with (–)-**10**

Host	Guest	Product	Yield (%)	Optical Purity (%)
(–)-**10**	**57a**	**58a**	17	68
(–)-**10**	**57b**	**58b**	30	67
(–)-**10**	**57c**	**59c**	69	97
(–)-**10**	**57d**	**58d**	25	53
(–)-**10**	**57e**	**58e**	87	100
(–)-**10**	**57f**	**58f**	56	100
(–)-**10**	**57g**	**58g**	42	100

Regioselective and enantioselective photodimerization of coumarin **60a** was achieved in the inclusion complexes with (*R,R*)-(–)-**8**, (*S,S*)-(–)-**3** and meso-**2**. (Table 9). For example, irradiation of the 1:1 inclusion compound of coumarin with (*R,R*)-(–)-**8** gave the anti-head-to-head dimer (–)-**61a** of 96%ee in 96% yield [25]. On the other hand 1:2 inclusion crystal of (*S,S*)-(–)-**3** with **60a** or the 1:2 inclusion crystal of meso-**2** with **60a** afforded the syn-head-to-head-dimer **62a** in 75% yield or anti-head-to-tail dimer (±)-**63a** in 94% yield, respectively [26]. Photodimerization of thiocoumarin **60b** is also

Table 9   Photodimerization of coumarin

Host	H:G	Irradiation time (h)	Product	Yield (%)
(*R,R*)-(–)-**8**	1:1	1	(–)-**61a** anti-head-to-head	96
(*S,S*)-(–)-**3**	1:2	5	**62a** syn-head-to-head	75
meso-**2**	1:2	4	(±)-**63a** anti-head-to-tail	94

Table 10 Photodimerization of thiocoumarin

Host	H:G	Irradiation time (h)	Product	Yield (%)
(R,R)-(−)-**8**	1:1	2	(+)-**61b** *anti-head-to-head*	73
(S,S)-(−)-**3**	1:2	4	**62b** *syn-head-to-head*	74
*meso-***2**	1:2	12	(±)-**63b** *anti-head-to-tail*	69

found to be controlled perfectly by using the host compounds, (R,R)-(−)-**9**, (S,S)-(−)-**3** and meso-**2** (Table 10).

60　a: X=O
　　　b: X=S

Cycloocta-2,4-dien-1-one **64** exists as an equilibrium mixture of the conformers **64a** and **64b** in solution, and conversion between these two enantiomeric forms is too fast to allow their isolation at room temperature. Photoreaction of **64** in pentane for 1 h gives racemic **65** in 10% yield along with polymers. When a solution of (−)-**3** and **64** was kept at room temperature for 12 h, a 3:2 inclusion complex of (−)-**3** with **64a** was obtained as colorless needles. Irradiation of the 3:2 complex of (−)-**3** with **64a** for 48 h gave (−)-**65** of 78% ee in 55% yield [27]. The chiral conformation of **64a** was determined by X-ray crystal structure analysis (Figure 14) [28].

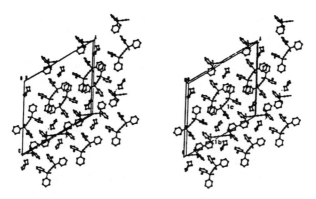

Figure 14   Stereoscopic view of the crystal structure of the 3:2 inclusion complex of (−)-**3** and **64**. Reprinted with permission from *J. Org. Chem.* **55** (1990) 4535. © American Chemical Society.

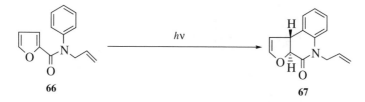

64a   **64b**   **65**

Enantioselective photocyclization of N-allylfuran-2-carboxanilide **66** in its inclusion crystals with the optically active host compound **9** was accomplished successfully [29]. More interestingly, (−)-**67** and (+)-**67** were obtained selectively upon photoirradiation of the 1:1 and 2:1 complexes of **66** with (−)-**9**, respectively. For example, photoirradiation of the powdered 1:1 complex of **66** with (−)-**9** gave (−)-**67** of 96% ee in 50% yield. On the other hand, similar irradiation of the 2:1 complex of **66** with (−)-**10** gave the other enantiomer (+)-**67** of 98% ee in 86% yield (Table 11).

66   **67**

Table 11   Photocyclization of **66** in the inclusion crystal with **9** or **10**

Host	H:G	Product		
		Compd	Yield (%)	% ee
(−)-**9**	1:1	(−)-**67**	50	96
(−)-**9**	2:1	(+)-**67**	86	98
(−)-**10**	2:1	(+)-**67**	77	98

The 1:1 complexation was achieved by mixing the host and guest in the solid state, and the inclusion complexation was followed by measurement of the CD spectra every 30 min in Nujol mull (Figure 15). For example, as the complexation between (–)-**10** and **66** proceeded, the (+)- and (–)-Cotton effects appeared at about 275 and 300 nm, respectively, and these absorptions increased until the complexation was completed after 2 h, although no absorption was present at the beginning of the mixing process.

Enantioselective photocyclization of 2-arylthio-3-methylcyclohexan-1-ones **68** to dihydrobenzothiophene derivatives **70** was also achieved in inclusion crystals using chiral host molecules [30]. Photoirradiation of the 1:1 inclusion crystals of **68g** with (–)-**9** as a water suspension gave the corresponding photocyclization product (+)-cis-**70g** of 82% ee in 83% yield. Similar photoirradiation of the inclusion crystals of **68a–f** with (–)-**9** afforded (+)-cis-**70b–f** of the optical purities listed in Table 12. X-ray crystal structural analysis of the 1:1 inclusion complex of **68g** with (–)-**9** showed the dihedral angle between the two average ring planes of the guest molecule to be about 74°. The photoreactive carbon C12 is 3.7 Å away from the target C3 on one side of the cyclohexenone ring plane, favoring the R-configuration at C3 (Figure 16). This assumption was ascertained by X-ray analysis of the photocyclization product (+)-**70g**. The chiral host molecule (–)-**9** showed neither strong absorption nor CD peaks in the 400–250 nm region. In contrast, the inclusion compound of prochiral molecules **68g** exhibited a rather strong CD spectrum (Figure 17). When cocrystallized with the opposite enantiomer, host (+)-**9**, an almost mirror-image solid-state CD spectrum was obtained.

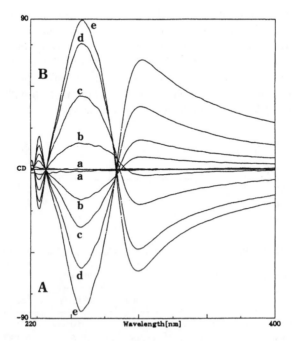

Figure 15   CD spectra of a 1:1 mixture of **66** with (–)-**9** (A) or (+)-**9** (B) after (a) 0 min, (b) 30 min, (c) 60 min, (d) 90 min and (e) 120 min in the Nujol mulls.

Table 12 Photocyclization of **68** in the inclusion crystals with **9**

Host	Guest	Product		
		Compd	Yield (%)	Optical purity (% ee)
(−)-**9**	**68a**	(+)-**70a**	90	32
(−)-**9**	**68b**	(+)-**70b**	75	60
(−)-**9**	**68c**	(+)-**70c**	92	62
(−)-**9**	**68d**	(+)-**70d**	94	65
(−)-**9**	**68e**	(+)-**70e**	85	77
(−)-**9**	**68f**	(+)-**70f**	83	59
(−)-**9**	**68g**	(+)-**70g**	83	82

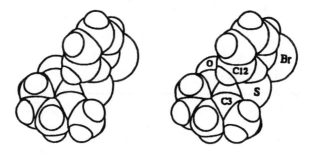

Figure 16 Molecular conformation of **68g** in the inclusion complex with (−)-**9**. Reprinted with permission from *J. Am. Chem. Soc.* **118** (1996) 11316. © American Chemical Society.

68 a: R=H   e: R=*o*-Cl   **69**   (+)-cis-**70**
       b: R=*p*-Me   f: R=*p*-Br
       c: R=*o*-Me   g: R=*o*-Br
       d: R=*p*-Cl

Host-guest inclusion complexes can be prepared by recrystallization of the host and guest compounds from a solvent. In some cases, however, the inclusion complex is not formed by this method. In such cases, mixing of powdered host and guest compounds in the absence of a solvent can give inclusion complexes [31]. For example, when the powdered 2:1 complex of (−)-**8** with MeOH, and N,N-dimethylphenylglyoxylamide **38a**, were mixed for 1 h using an agate mortar and pestle, the mixture solidified to give a 2:1 inclusion complex crystal of (−)-**8** and **38a**. Upon formation of this complex, the ν(OH) of **8** (3550 and 3380 cm^{-1}) shifted to lower wavenumbers (3310 and 3250 cm^{-1}) due to hydrogen bond formation between the OH group of **8** and the CO group of **38a** in the

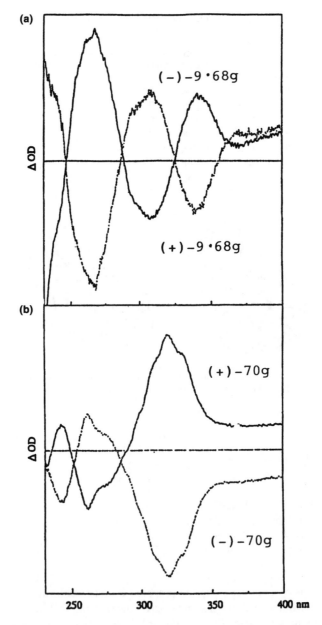

Figure 17 CD spectra of (a) inclusion complexes of **68g** with (−)-**9** and (+)-**9** and of (b) (+)-**70g** and (−)-**70g**. Reprinted with permission from *J. Am. Chem. Soc.* **118** (1996) 11316. © American Chemical Society.

complex. Irradiation of the powdered 2:1 complex of (−)-**8** and **38a** as a suspension in water for 10 h gave (+)-**39a** of 61% ee in 70% yield. On the other hand, (−)-**9** formed a 1:1 inclusion complex with **38a** either by mixing in the solid state or by recrystallization from diethyl ether. Photoirradiation of these inclusion complexes in a water suspension

Table 13 Formation of optically active **39a** and **40a** by irradiation for 10 h of the inclusion complexes of chiral host compounds and **38a** prepared by mixing in the solid state or recrystallization from toluene

Host	H:G	Product via mixing			Product via recrystallization		
		Compd	Yield (%)	% ee	Compd	Yield (%)	% ee
(–)-**8**	2:1	(+)-**39a**	70	61	—		
(–)-**9**	1:1	(–)-**39a**	29	82	(–)-**39a**	47	79
		(–)-**40a**	35	45	(–)-**40a**	23	42
(–)-**10**	2:1	(+)-**39a**	48	41	(–)-**39a**	39	85

gave (–)-**39a** and (–)-**40a** in the optical and chemical yields shown in Table 13. (–)-**10** also formed a 2:1 inclusion complex with **38a** by either the mixing or recrystallization method; however, photoirradiation of these two complexes in an aqueous suspension gave (+)- or (–)-**39a**, respectively. It is very interesting that the direction of the chiral arrangement of **38a** in the complexes with (–)-**10** changes depending on the preparation method used.

Most interestingly, interconversion between the molecular arrangement of **38e** in the (–)- and (+)-forms easily occurs by complexation with the chiral host (–)-**10** in the solid state [32]. By mixing the powdered 1:1 MeOH complex of (–)-**10** and powdered (–)-crystals of **38e**, 1:1 inclusion crystals of (–)-**10** and (+)-**38e** were formed. The (+)-arrangement of **38e** in the above inclusion crystals can be proven by its photoirradiation in a water suspension for 5 h which gives (+)-**39e** of 76% ee in 72% yield, and its absolute configuration was determined by X-ray analysis. The chirality of **38e** was also easily detected by measurement of the CD spectra as Nujol mulls as shown in Figure 18. The conversion of (S)-(–)-**38e** into (R)-(+)-**38e** during the complexation in the solid state was observed by continuous measurement of the CD spectra as Nujol mulls. As the complexation proceeds, strong CD spectra showing a (–)-Cotton effect due to (S)-(–)-**38e** are gradually converted into spectra showing the (+)-Cotton effect of (R)-(+)-**38e** (Figure 19). This reaction was also monitored by recording the IR spectra as Nujol mulls (Figure 20).

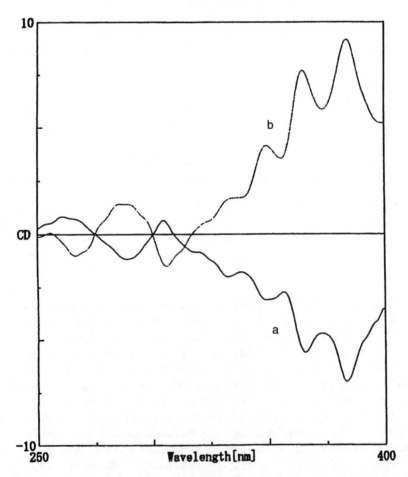

Figure 18 CD spectra of (a) (S)-(−)- and (b) (R)-(+)-**38e** crystal. Reprinted with permission from *J. Org. Chem.* **62** (1997) 9262. © American Chemical Society.

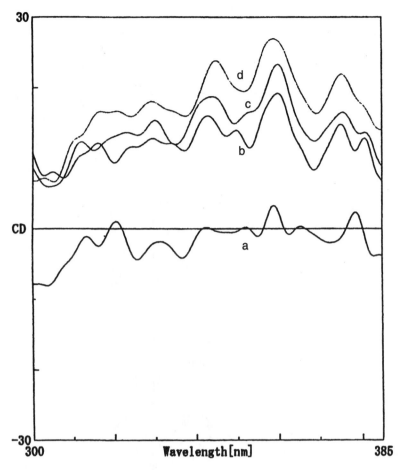

Figure 19 CD spectra of a mixture of (S)-(–)-**38e** and a 2:1 complex of (–)-**9** with MeOH in Nujol mulls after (a) 0 min, (b) 30 min, (c) 60 min and (d) 180 min. Reprinted with permission from *J. Org. Chem.* **62** (1997) 9262. © American Chemical Society.

Mixing of powdered thiocoumarin **60b** and the optically active host compound (R,R)-(–)-**9** in the solid state gave their 1:1 complex, in which the former molecules are arranged in a chiral form. Photoirradiation of these 1:1 inclusion crystals in the solid state gave the optically active anti-head-to-head dimer (+)-**61b** of 100% ee in 73% yield. The inclusion complexation of **60b** and (–)-**9** was followed by continuous measurement of the CD spectra as Nujol mulls [33]. Although a mixture of **60b** and (–)-**9** as Nujol did not show a clear CD absorption initially, induced CD absorption of **60b** appeared and increased as the complexation proceeded (Figure 21). The enantioselective photodimerization reaction of **60b** was also followed by CD spectral measurement. The CD absorptions at 260, 320, and 370 nm due to **60b** in the inclusion crystal disappeared and new CD absorptions of 259 at 270 and 330 nm appeared after 5 min photoirradiation (Figure 22).

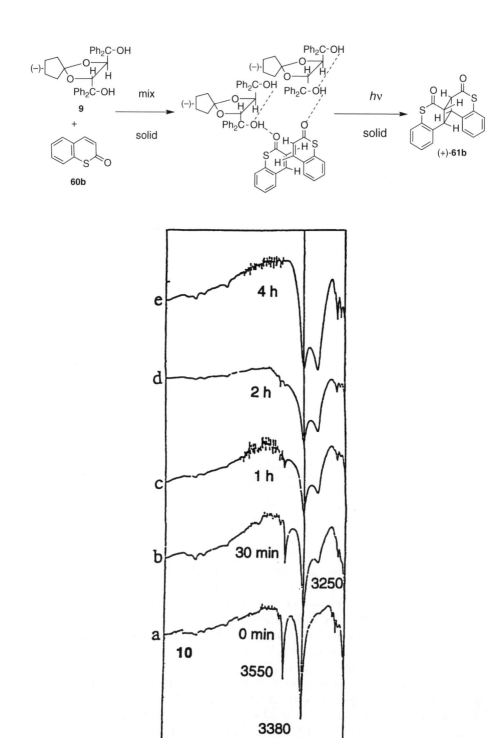

Figure 20 IR spectra of a mixture of (S)-(–)-**38e** and a 2:1 complex of (–)-**9** with MeOH in Nujol mulls.

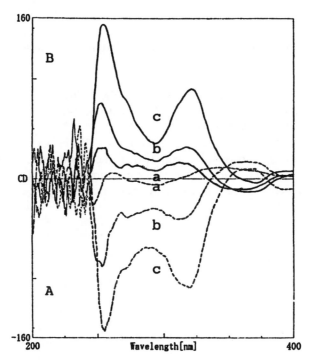

Figure 21 CD spectra of a mixture of (−)-**9** (A) or (+)-**9** (B), **60b** and liquid paraffin after (a) 0 min, (b) 30 min and (c) 60 min.

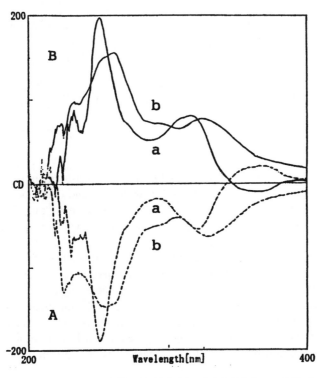

Figure 22 CD spectra of a 1:1 complex of **60b** with (−)-**9** (A) or (+)-**9** (a) before and (b) after photoirradiation for 5 min.

4. Single-Crystal-to-Single-Crystal Photoreaction in Host-Guest Inclusion Crystals

In some cases, enantioselective photoreactions in the inclusion crystals with optically active host compounds proceed in a single-crystal-to-single-crystal manner without collapse of the crystalline lattice.

For example, irradiation of the 2:2 inclusion crystal (71) of (–)-8 with 60a in the solid state with a 400-W high-pressure Hg lamp (Pyrex filter, room temperature, 4 h) gave a 2:1 complex (72) of (–)-8 with (–)-61a. After photoirradiation, the crystals were still clear and the reaction proceeded in a single-crystal-to-single-crystal manner throughout the reaction [34].

The photodimerization of 71 to 72 was followed by measurement of the CD spectra as Nujol mulls. The 1:1 complex (71) of 60a with (+)- and (–)-8 showed CD spectra with a mirror-imaged relation (Figure 23). After photoirradiation, the CD absorptions of 71 at 225, 275, 300 and 330 nm disappeared, and a new CD absorption due to 72 at 240 nm appeared.

The single-crystal-to-single-crystal nature and the steric course of the photodimerization of coumarin 60a to (–)-anti-head-to-head dimer 61a in the inclusion complex (71) were investigated by X-ray crystallographic analysis and X-ray powder diffraction spectroscopy. Crystal data are listed in Table 14. X-ray crystallographic analysis showed that two molecules of 60a were arranged to form a hydrogen bond between the C40=O6 of 60a and the O4-H of 8 in the direction which gave the anti-head-to-head dimer (61a) by photodimerization and the molecular aggregation with 3.59 and 3.42 Å was short

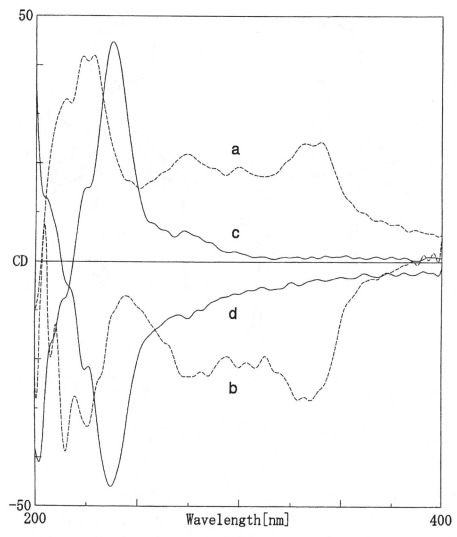

Figure 23 CD spectra in Nujol mulls: (a) a 1:1 complex of **60a** with (+)-**8**; (b) a 1:1 complex of **60a** with (–)-**8**; (c) a 2:1 complex of (+)-**8** with (+)-**61a**; (d) a 2:1 complex of (–)-**8** with (–)-**61a**. Reprinted from *Tetrahedron* **56** (2000) 6853–6865, with permission from Elsevier Science.

enough, which should react easily and topochemically (Figure 24). After photoirradiation, the bond distances of the cyclobutane ring connecting C38=C38* and C39=C39* are 1.6 and 1.57 Å, respectively (Figure 25).

The crystal-to-crystal nature of this reaction was also confirmed by X-ray powder diffraction spectroscopy. In the inclusion compounds **71** and **72**, the peaks at $2\theta = 8.90°$, $9.92°$ disappeared, and new peaks at $2\theta = 5.36°$, $8.46°$, $10.78°$ appeared during UV irradiation. After irradiation for 4 h, the original crystal structure was converted almost completely to the new structure (Figure 26).

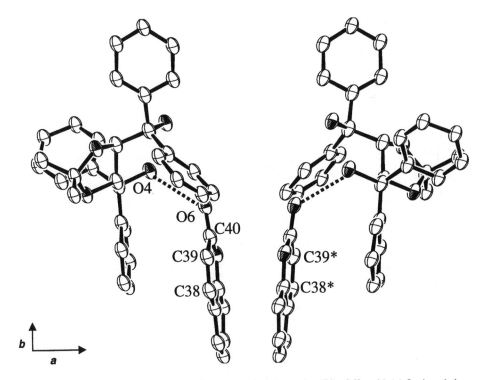

Figure 24   ORTEP drawing of the molecular structure of a 1:1 complex (**71**) of **60a** with (–)-**8**: viewed along the *a*-axis. The hydrogen bonds are shown by dotted lines. Reprinted from *Tetrahedron* **56** (2000) 6853–6865, with permission from Elsevier Science.

Table 14   Crystallographic data for a 1:1 complex (**71**) of **60a** with (–)-**8**, a 2:1 complex (**72**) of (–)-**61a**, a 1:1 complex (**73**) of **60b** with (–)-**9**, and a 2:1 complex (**74**) of (–)-**9** with (+)-**61b**

Compound	**71**	**72**	**73**	**74**
Formula	$C_{40}H_{36}O_6$	$C_{40}H_{36}O_6$	$C_{42}H_{38}O_5S_1$	$C_{42}H_{38}O_5S_1$
Crystal system	monoclinic	monoclinic	monoclinic	monoclinic
Space group	$C2$	$C2$	$P2_1$	$P2_1$
$a/\text{Å}$	35.59 (4)	32.80 (3)	10.235 (2)	10.371 (3)
$b/\text{Å}$	9.489 (3)	9.467 (3)	35.78 (1)	34.70 (2)
$c/\text{Å}$	10.03 (1)	10.360 (4)	9.422 (2)	9.414 (3)
$d/\text{Å}$	102.70 (4)	100.27 (7)	91.00 (2)	91.38 (2)
$V/\text{Å}^3$	3305 (4)	3164 (2)	3449 (1)	3387 (2)
Z	4	4	4	4
D calc	1.23	1.29	1.26	1.28
R	0.101	0.114	0.065	0.078
Temp/°C	rt	rt	–50	–50

The difference between the molecular structures of **71** and **72** is shown in Figure 27. As seen in Figure 25, the cyclobutane ring forms approximately along the *a*-axis. This corresponds well to the anisotropic changes in the lattice constants. There is not so much movement in the host molecule, though the molecular structure of the guest molecule is

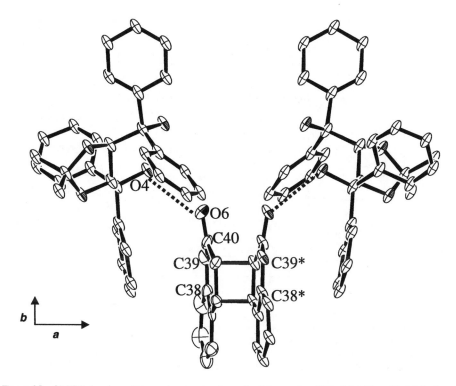

Figure 25 ORTEP drawing of the molecular structure of a 2:1 complex (**72**) of (–)-**61a** with (–)-**8** viewed along the *a*-axis. The hydrogen bonds are shown by dotted lines. Reprinted from *Tetrahedron* **56** (2000) 6853–6865, with permission from Elsevier Science.

changed largely with the formation of a cyclobutane ring. The host molecules play an important role in maintaining the crystal lattice during the reaction.

The enantioselective photodimerization of thiocoumarin (**60b**) to optically pure (+)-*anti*-head-to-head dimer (**61b**) in the 1:1 inclusion complex (**73**) of **60b** with (–)-**9** was also found to proceed in a single-crystal-to-single-crystal manner. Photoirradiation of **73** in the solid state (400-W high-pressure Hg lamp, Pyrex filter, room temperature, 2 h) gave a 2:1 complex (**74**) of (–)-**9** with (+)-**60b**, quantitatively.

Chiral transformation of the achiral molecule **60b** in the inclusion complex (**73**) can easily be detected by CD spectral measurement of its Nujol mull. The 1:1 complex of **60a** with (+)- and (–)-**9** showed CD spectra with a mirror-imaged relation (Figure 28). After photoirradiation, the CD absorptions of **73** at 260 and 320 nm disappeared, and new CD absorptions due to **74** at 270 and 330 nm appeared.

The single-crystal-to-single-crystal nature and the steric course of the photodimerization of thiocoumarin **60b** to (+)-anti-head-to-head dimer **61a** in the inclusion complex **73** were also elucidated by X-ray crystallographic analysis and X-ray powder diffraction spectroscopy. Two molecules of **60b** in a 1:1 inclusion complex **73** are related to a pseudo two-fold axis along the *c*-axis (Figure 29). C75=O9, C84=O10 of **60b** and the O4=H, O7=H of **9** form a hydrogen bond in the direction which gives the *anti*-head-to-head

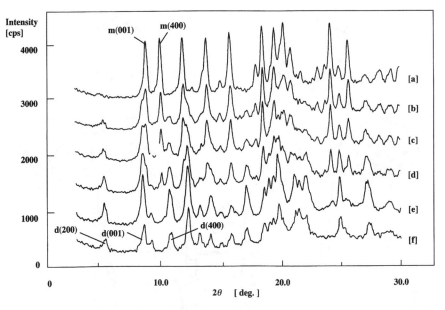

Figure 26 X-ray powder diffraction patterns of (a) a 1:1 complex (**71**) of **60a** with (–)-**8**(b) after UV irradiation for 10 min (c) after UV irradiation for 30 min (d) after UV irradiation for 90 min (e) after UV irradiation for 4 h (f) a 2:1 complex (**72**) of (–)-**61a** with (–)-**8**; m(001) and m(400) in (a) are the mirror indices for complex (**71**). d(200), d (001) and d(400) in (f) are the mirror indices for complex (**72**).

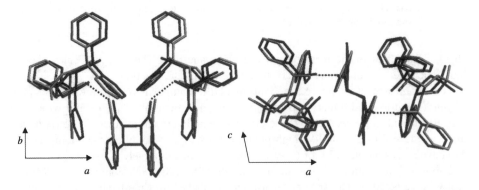

Figure 27 Superposition of monomer and dimer conformation: A 1:1 complex (**71**) of **60a** with (–)-**8** (in pink) and a 2:1 complex (**72**) of (–)-**61a** with (–)-**8** (in brown), left is viewed along the *c*-axis and right is viewed along the *b*-axis. B left is viewed along the *a*-axis and right is viewed along the *c*-axis. The hydrogen bonds are shown by dotted lines. Reprinted from *Tetrahedron* **56** (2000) 6853–6865, with permission from Elsevier Science.

dimer (**61b**). The distance between the two ethylenic double bonds is short enough (3.73 and 3.41 Å) to react easily and topochemically. After photoirradiation, the bond distances of the cyclobutane ring connecting C74=C83 and C73=C82 are 1.60 and 1.60 Å, respectively. The distance of atoms which formed the hydrogen bond

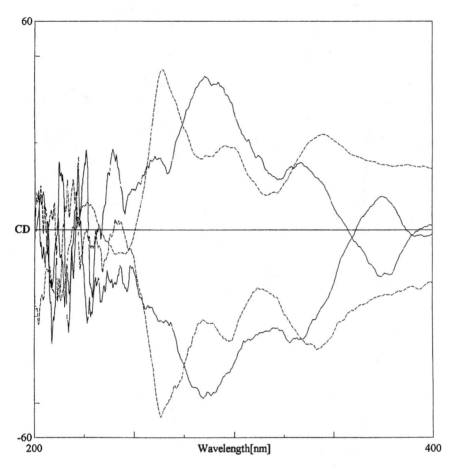

Figure 28 CD spectra in Nujol mulls: (a) a 1:1 complex of **60b** with (+)-**9**; (b) a 1:1 complex of **60b** with (–)-**9**; (c) a 2:1 complex of (+)-**9** with (–)-**61b**; (d) a 2:1 complex of (–)-**9** with (+)-**61b**. Reprinted from *Tetrahedron* **56** (2000) 6853–6865, with permission from Elsevier Science.

(O4=H···O9=C75, O7=H···O10=C84) are 2.81 and 2.77 Å in a 1:1 inclusion complex **74** of **1b** with (–)-**2b**, respectively (Figure 30). From the superposition of the molecular structure (Figure 31), the thiocoumarin ring and phenyl group are stacked to synchronously transform in the case of **73** to **74**, and little rotation occurs in the crystalline state.

In the X-ray powder patterns of **73** and **74**, the peaks at $2\theta=8.64°$, $9.42°$ disappeared, and new peaks at $2\theta=5.20°$, $8.76°$, $10.32°$ appeared during UV irradiation. After irradiation for 4 h, the original crystal structure was transformed almost completely to the new structure (Figure 32).

A solution of (–)-**4** and **2** in ether-hexane (1:1) was kept at room temperature for 6 h to give a 1:2 complex (**16**) of (–)-**4** and **2** as colorless prisms in 91% yield [35]. Photoirradiation of single-crystals of **16** (mp 90–9 °C) with a 400-W high-pressure

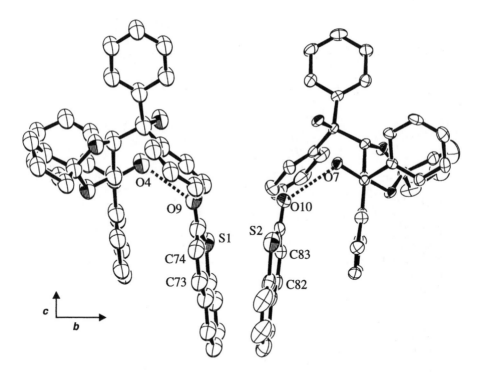

Figure 29 ORTEP drawing of the molecular structure of a 1:1 complex (**73**) of **60b** with (+)-**9** viewed along the *a*-axis. The hydrogen bonds are shown by dotted lines. Reprinted from *Tetrahedron* **56** (2000) 6853–6865, with permission from Elsevier Science.

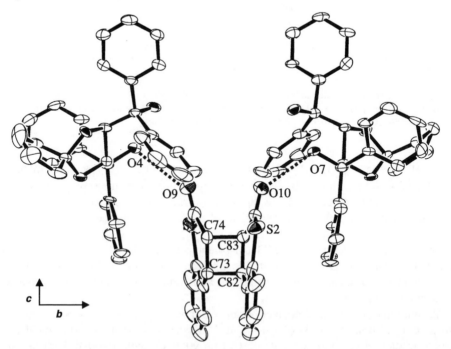

Figure 30 ORTEP drawing of the molecular structure of a 2:1 complex (**74**) of (–)-**9** with (+)-**61b** viewed along the *a*-axis. The hydrogen bonds are shown by dotted lines. Reprinted from *Tetrahedron* **56** (2000) 6853–6865, with permission from Elsevier Science.

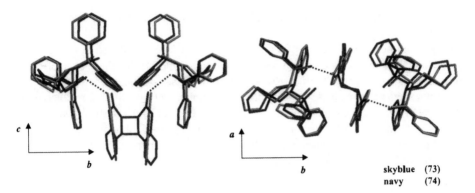

Figure 31 Superposition of monomer and dimer conformation: A 1:1 complex (**73**) of **60b** with (–)-**9** (in skyblue), a 2:1 complex (**74**) of (–)-**9** with (+)-**61b**(in navy), left is viewed along the *a*-axis and right is viewed along the *c*-axis. The hydrogen bonds are shown by dotted lines. Reprinted from *Tetrahedron* **56** (2000) 6853–6865, with permission from Elsevier Science.

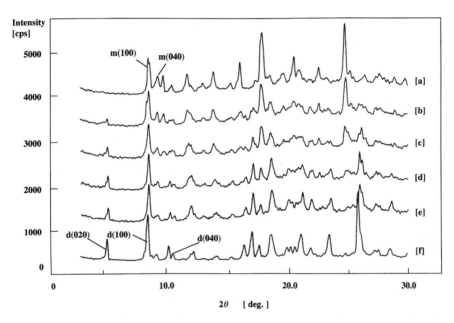

Figure 32 X-ray powder diffraction patterns of (a) a 1:1 complex (**73**) of (–)-**9** with **60b** (b) after UV irradiation for 10 min (c) after UV irradiation for 20 min (d) after UV irradiation for 30 min (e) a 2:1 complex (**74**) of (–)-**9** with (+)-**61b**; m (100) and m (040) in [a] are the mirror indices for complex (**73**). *d* (020), *d* (100) and *d* (040)in [f] are the mirror indices for complex (**74**). Reprinted from *Tetrahedron* **56** (2000) 6853–6865, with permission from Elsevier Science.

Hg-lamp through a Pyrex filter for 24 h gave single crystals of a 2:1 complex **17** (mp 79–84 °C) of **4** with (–)-**10**, which upon distillation *in vacuo* gave (–)-**10** of 46% ee in 75% yield. When the reaction was carried out at −78 °C in the solid state for 30 h, (–)-**10** of 58% ee was obtained in 55% yield. After photoirradiation, the crystal was still clear and the reaction proceeded in a single-crystal-to-single-crystal manner throughout the reaction.

The photodimerization reaction of cyclohex-2-enone **2** in the inclusion crystal with (−)-**4** also occurs in a single-crystal-to-single-crystal manner [36]. Photoirradiation of single-crystals of **77** (mp 90–95 °C) with a 400-W high-pressure Hg-lamp through a Pyrex filter for 24 h gave single-crystals of a 2:1 complex **78** (mp 79–84 °C), which upon distillation *in vacuo* gave (−)-**76** of 46% ee in 75% yield. When the reaction was carried out at −78 °C in the solid state for 30 h, (−)-**76** of 58% ee was obtained in 55% yield. After photoirradiation, the crystal was still clear and the reaction proceeded in a single-crystal-to-single-crystal manner throughout the reaction.

The single-crystal-to-single-crystal nature of the reaction was confirmed by X-ray single crystal structure analysis. The host molecules **4** in the complex crystal **77** stack along the *a*-axis, to make the inclusion columns, where the guest molecules **2A** and **2B** are alternately located (Figure 33). The C4=C5 bond in **2A** makes a short contact with the C7=C8 in **2B**(C4=C8=3.99, C5=C7=4.16 Å) giving the reaction product of (−)-**2**. The distances between the π orbital lobe apexes are 1.76 Å for C4 and C8, and 1.56 Å for C5 and C7 atoms. The arrangement of the host molecules in the complex crystal **78** was similar to that of the complex crystal **77**(Figure 34). The photoreaction product could be located as (−)-**76** in the inclusion column with a considerably distorted shape from the ideal one and a large apparent thermal motion perpendicular to the molecular plane. However, there should be (+)-**76** as a minor product in the complex crystal **78**, since the reaction product consists of (−)-**76** and (+)-**76** in about 3:1 ratio.

A comparison of the crystal structure of **77** with that of **78** shows that hosts can move parallel to the *a*-axis, but in the reverse direction, giving the close approach (about 1.4 Å) of **2A** and **2B**, respectively. The mirror product ((+)-**76**) will be given if the molecule **2A** rotates toward **2B*** along the axis joining C3 and C6 atoms, and the C4=C5 bond of **2A** approaches to react with C7*=C8*. The long distances between C4=C5 and C7*=C8* bonds (C4–C8*=4.82 and C5–C7*=4.46 Å) makes this approach difficult, resulting in the formation of (+)-**76** as a minor product.

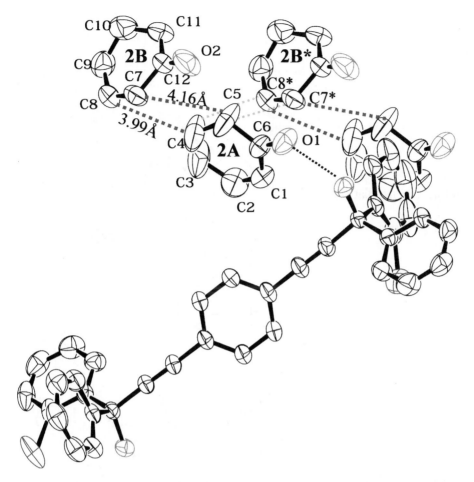

Figure 33 A view of the arrangement of the guest and the host molecules in the complex crystal **77**. The contacts of the C4–C5 bond with the C7–C8 and C7*–C8* bonds are shown by dotted lines. Reprinted from *Tetrahedron* **56** (2000) 6853–6865, with permission from Elsevier Science.

5. Enantioselective Photoreaction in Chiral Crystals

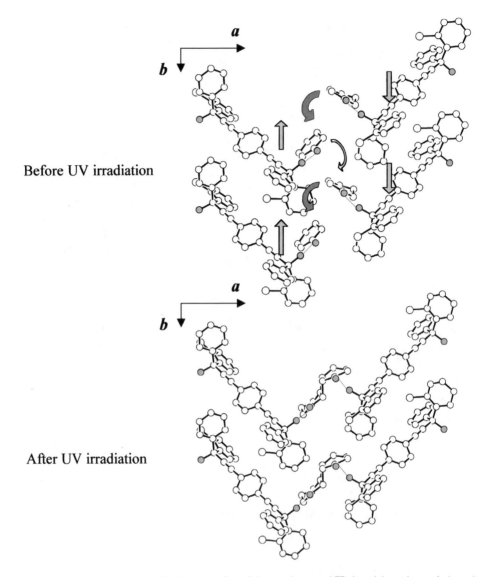

Before UV irradiation

After UV irradiation

Figure 34 Molecular packing and hydrogen bonding of the complex crystal **77** viewed down the *c*-axis (upper) and molecular packing and hydrogen bonding of the complex crystal **78** viewed down the *c*-axis (below). Reprinted from *Tetrahedron* **56** (2000) 6853–6865, with permission from Elsevier Science.

Several successful examples of the so-called 'absolute' asymmetric synthesis by using chiral crystals of achiral molecules in the absence of any external chiral source have been found recently, and they provide an attractive and new method for asymmetric synthesis. It is also highly relevant in relation to the origin of optically active compounds on the earth.

N,N-Diisopropylbenzoylformamide **38f** forms chiral crystals which give the optically active β-lactam derivative **39f** of high optical purity upon photoirradiation in the solid state [37]. Recrystallization of **38f** from benzene afforded colorless prisms. Crystals of **38f** which give (+)- and (–)-**39f** on photocyclization have been tentatively identified as the (+)- and (–)-crystal forms of **38f**, respectively. Large amounts of the (+)- and (–)-crystals of **38f** can easily be prepared by seeding with finely powdered (+)- and (–)-crystals, respectively, during recrystallization of **38f**. Irradiation of (+)-crystals of **38f** with a 400-W high-pressure Hg-lamp for 40 h at room temperature gives (+)-**39f** of 93% ee in 74% yield. Irradiation of (–)-crystals of **38f** under the same conditions gives (–)-**39f** of 93% ee in 75% yield. X-ray crystal structural analysis of a (+)-crystal of **38f** shows that molecules of **38f** are twisted around the CO=CO bond in the chiral crystal and hence photocyclization gives optically active **39f** (Figure 35).

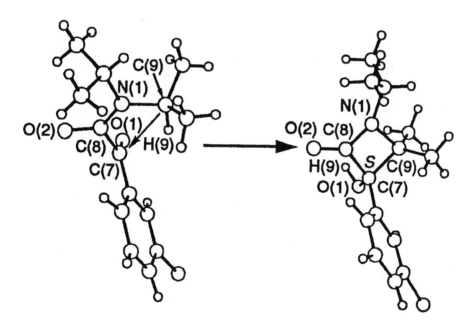

79 → **80**

a: R = *m*-Me   e: R = *p*-Cl   i: R = *o*-Br
b: R = *m*-Cl   f: R = *p*-Br   j: R = 3,5-Me$_2$
c: R = *m*-Br   g: R = *o*-Me   k: R = 3,4-Me$_2$
d: R = *p*-Me   h: R = *o*-Cl

Figure 35   The process of enantioselective photocyclization **38b** to **39b**. Reprinted from *JCS Perkin Trans 2* (1996) 64, by permission of the Royal Society of Chemistry.

Table 15 Photocyclization of substituted phenyglylamides in their crystals

Phenylgloxylamide	Irradiation time (h)	Yield (%)	Optical purity (% ee)
79a	7	63	91
79b	10	75	100
79c	5	97	96
79d	5	60	0
79e	5	50	0
79f	12	65	0
79g	10	54	9
79h	24	42	0
79i	24	48	0
79j	5	74	0
79k	5	62	54

The formation of such chiral crystals depends on the substituent on the benzene ring of the phenylglyoxylamide [38]. All the meta-substituted N,N-diisopropylphenylglyoxy-lamides **79a–c** tested formed chiral crystals, and their photoirradiation in the solid state gave the optically active β-lactams **80a**, **80b** and **80c**, respectively. The optical purity and yield for these products are shown in Table 15. However, all the para-substituted isomers tested, **79d–f**, did not form chiral crystals and these gave rac-β-lactams upon photoirradiation (Table 15). On the other hand, although ortho-methyl substituted phenylglyoxylamide **79g** formed chiral crystals and gave optically active **80g** on irradia-tion, the ortho-chloro-**79h** and the ortho-bromo-substituted compound **79i** did not form chiral crystals and gave rac-**80h** and **80i**, respectively. However, meta-substituted phenylglyoxylamide **79j** did not form chiral crystals and its photoreaction gave rac-**80j**, although the meta, para-disubstituted one **79k** formed chiral crystals and gave optically active **80k** (Table 15). When the isopropyl group of **79** is replaced with other alkyl groups such as benzyl or cyclohexyl, no chiral crystals are formed, and photoirradiation of these derivatives gave rac-β-lactams. It is clear that the isopropyl group may play an important role in the formation of these chiral crystals.

Table 16 Crystal data of **79b**, **79e** and **79h**

Crystal data	Cl—⟨benzene⟩—COCON*i*Pr$_2$ **79h**	Cl—⟨benzene⟩—COCON*i*Pr$_2$ **79b**	Cl—⟨benzene⟩—COCON*i*Pr$_2$ **79e**
Crystal system	Monoclinic	Orthorhombic	Orthohombic
Space group	P2$_1$/n	P2$_1$2$_1$2$_1$	Pbca
Z	4	4	8
a(Å)	9.582	12.800	15.177
b(Å)	21.494	14.941	15.558
c(Å)	7.591	7.611	12.555
β(°)	111.40	–	–
V(Å^3)	1455.6	1455.6	2964.6

In order to elucidate the process of asymmetric photocyclization, the crystal structures of **79b**, **79e** and **79h** were determined (Table 16). When the crystal of **79e** is irradiated with UV light, a radical is probably produced at the oxygen, O1, of the carbonyl group. The oxygen radical produced should abstract the hydrogen atom H9 of the isopropyl group to form a hydroxy group. The radical produced at C9 due to this abstraction of H9 would then attack the C7 atom to form a C=C single bond. The absolute structures of **79b** and **80b** in Figure 35 are in agreement with those determined with anomalous dispersion terms for the chlorine atoms. If the radical C9 attacks the atom C7 from the si-face of the carbonyl group as shown in Figure 35, the β-lactam produced should have the S configuration, which is in agreement with the experimental result.

The achiral molecule 3,4-bis(diphenylmethylene)succinimide **81** was found to arrange itself in a chiral form in the crystalline state [39]. The chirality of **81** was frozen by photo-irradiation in the solid state to give the optically active photocyclization product **83**. Recrystallization of **81** from acetone formed chiral crystals as orange hexagonal plates (A, converts to C by heating at 260 °C) and two types of racemic crystals as orange rectangular plates (B, mp 302 °C) and yellow rectangular plates (C, mp 297 °C) (Figure 36). The chirality of the crystal A can be easily detected by measurement of its CD spectra as Nujol mulls. The crystal A exhibits strong CD absorptions at around 250 and 330 nm (Figure 37), while the crystal types B and C do not show any CD absorption in these regions.

153

Figure 36 The photographs of three crystal modifications of **81**. Reprinted with permission from *Chem. Rev.* (2000) 1069. © American Chemical Society.

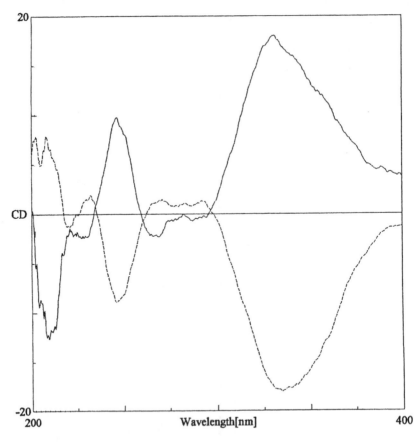

Figure 37 CD spectra of two enantiomeric crystal of **81A**. Reprinted with permission from *Chem. Rev.* (2000) 1069. © American Chemical Society.

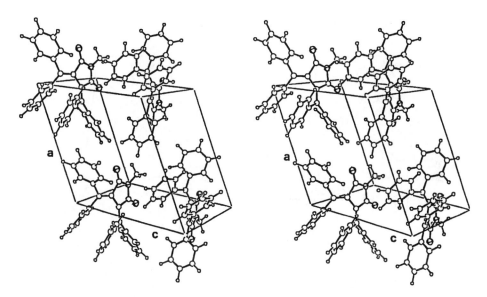

Figure 38   Stereoview of the molecular packing of **81A**.

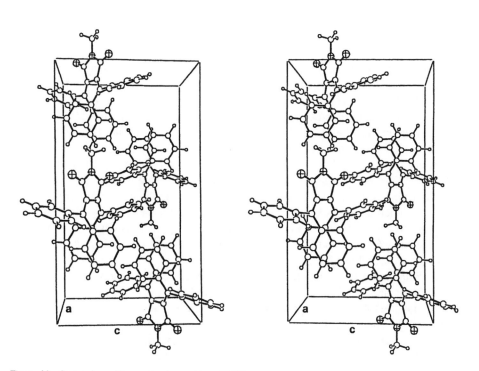

Figure 39   Stereoview of the molecular packing of **81B**.

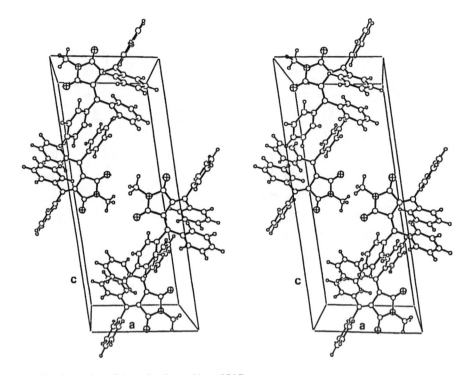

Figure 40 Stereoview of the molecular packing of **81C**.

Irradiation of powdered (+)-A crystals using a 100-W high-pressure Hg-lamp for 50 h gave (+)-**83** of 64% ee in quantitative yield. Similar irradiation of (–)-A crystals gave (–)-**83**. However, photoirradiation of crystals B and C gave racemic **83**. This enantioselective photoconversion consists of two steps, the conrotatory ring closure of **81** to the intermediate **82**, then a 1,5-hydrogen shift to give the product **83**. X-ray crystal structure analysis disclosed that crystal A consists of **81** molecules in either a purely right-handed or a purely left-handed helical configuration (Figure 38). Crystals B and C contain equal amounts of the two configurations, and hence are racemic (Figures 39 and 40). The phenyl rings a and a' are almost parallel to one another and overlap significantly: the dihedral angles between these rings are within the range 9–12°. The corresponding distances between the unsaturated carbon sites which join during the photoreaction, C6⋯C30 and C14⋯C22, are 3.38 and 3.27 Å in A, 3.35 Å in B, and 3.35 and 3.34 Å in C (Table 17).

Transformation between the different polymorphs can easily be accomplished both in the solid state and in solution. Addition of one piece of (–)-A crystal during the recrystallization of (+)-A crystal (50 mg) from acetone gave (–)-A (23 mg), B (4 mg) and C crystal (8 mg). The thermal racemization of A to C occurs before melting, by heating to 260 °C on a hot plate. During this process the color changes from orange to yellow, spreads dramatically from one end of the crystal to the other with retention of the crystal morphology (Figure 41).

Table 17 Crystal data and selected structural parameters of three polymorphic crystals of **81**

Crystal data	A	B	C
Crystal system	Monoclinic	Orthorhombic	Monoclinic
Space group	$P2_1$	*Pbcn*	$P2_1/n$
a (Å)	11.640	9.964	9.485
b (Å)	9.257	20.181	11.014
c (Å)	12.103	11.622	22.945
Intramolecular nonbonding distances (Å)			
$C(6)\cdots C(30)$	3.381	3.353	3.348
$C(14)\cdots C(22)$	3.273	3.353	3.338
Torsion angles (°)			
$C(6)-C(5)-C(21)-C(22)$	38.8	34.7	43.2
$C(4)-C(3)-C(5)-C(6)$	19.6	13.7	20.2
$C(20)-C(19)-C(21)-C(22)$	14.3	13.7	20.9

Figure 41   The photographs of thermal conversions of **41a** and **41b** to **41c**.

References

1. D. RABINOVICH, *J. Chem. Soc. B* (1970) 11.
2. K. OHKUMA, S. KASHINO and M. HAIDA. *Bull. Chem. Soc. Jpn.* **46** (1972) 627.
3. H. STOBBE and K. BRENER. *J. Prakt. Chem.* **123** (1929) 241.
4. K. TANAKA and F. TODA. *J. Chem. Soc., Chem. Commun.* (1983) 593; M. KAFTORY, K. TANAKA and F. TODA. *J. Org. Chem.* **50** (1985) 2154.
5. F. TODA, K. TANAKA and M. KATO. *J. Chem. Soc., Perkin Trans.* **1** (1998) 1315.
6. K. TANAKA and F. TODA. *Nippon Kagakukaishi* (1984) 141; M. KUZUYA, A. NOGUCHI, N. YOKOTA, T. OKUDA, F. TODA and K. TANAKA. *Nippon Kagakukaishi* (1986) 1746; T. FUJIWARA, N. TANAKA, K. TANAKA and F. TODA. *J. Chem. Soc. Perkin Trans.* **1** (1989) 663.
7. F. TODA and K. TANAKA. *J. Chem. Soc., Chem. Commun.* (1986) 1429.; F. TODA, K. TANAKA and M. YAGI. *Tetrahedron* **43** (1987) 1493.
8. M. KAFTORY, M. YAGI, K. TANAKA and F. TODA. *J. Org. Chem.* **53** (1988) 4391.
9. F. TODA and K. TANAKA. *Tetrahedron Lett.* **29** (1988) 4299; T. FUJIWARA, N. TANAKA, K. TANAKA and F. TODA. *J. Chem. Soc., Perkin Trans.* **1** (1989) 663.
10. F. TODA and K. TANAKA. *Chem Lett.* (1987) 2283; F. TODA, K. TANAKA and T. MAK. *C. W. Chem. Lett.* (1989) 1329.
11. H. AOYAMA, T. HASEGAWA and Y. OMOTE. *J. Am. Chem. Soc.* **101** (1979) 5343.
12. F. TODA, K. TANAKA, M. YAGI, Z. STEIN and I. GOLDBERG. *J. Chem. Soc. Perkin Trans.* **1** (1990) 1215.

13. F. TODA, K. TANAKA and M. YAGI. *Tetrahedron* **43** (1987) 1493; M. KAFTORY, M. YAGI, K. TANAKA and F. TODA. *J. Org. Chem.* **53** (1988) 4391.

14. F. TODA, H. MIYAMOTO and R. MATSUKAWA. *J. Chem. Soc., Perkin Trans* **1** (1992) 1461; D. HASHIZUME, H. UEKUSA, Y. OHASHI, R. MATSUGAWA, H. MIYAMOTO and F. TODA. *Bull. Chem. Soc. Jpn.* **67** (1994) 985.

15. F. TODA, K. TANAKA, O. KAKINOKI and T. KAWAKAMI. *J. Org. Chem.* **58** (1993) 3783.

16. D. HASHIZUME, Y. OHASHI, K. TANAKA and F. TODA. *Bull. Chem. Soc. Jpn.* **67** (1994) 2383.

17. O. L. CHAPMAN and W. R. ADAMS. *J. Am. Chem. Soc.* **90** (1968) 2333; I. NINOMIYA and T. NAITO. *"The Alkaloids"* **22** (Academic Press, San Diego, 1983) p. 189.

18. I. NINOMIYA, S. YAMAUCHI, T. KIGUCHI, A. SHINOHARA and T. NAITO. *J. Chem. Soc., Perkin Trans* **1**. (1974) 1747.

19. K. TANAKA, O. KAKINOKI and F. TODA. *J. Chem. Soc., Chem. Commun.* (1992) 1053.

20. S. OHBA, H. HOSOMI, K. TANAKA, H. MIYAMOTO and F. TODA. *Bull. Chem. Soc. Jpn.* **73** (2000) 2075.

21. H. HOSOMI, S. OHBA, K. TANAKA and F. TODA. *J. Am. Chem. Soc.* **122** (2000) 1818.

22. F. TODA, H. MIYAMOTO, K. TAKEDA, R. MATSUGAWA and N. MARUYAMA. *J. Org. Chem. Soc.* **58** (1993) 6208; S. AKUTSU, I. MIYAHARA, K. HIROTSU, H. MIYAMOTO, N. MARUYAMA, S. KIKUCHI and F. TODA. *Mol. Cryst. Liq. Cryst.* **277** (1996) 87.

23. F. TODA, H. MIYAMOTO, M. INOUE, S. YASAKA and I. MATIJASIC. *J. Org. Chem.* **65** (2000) 2728.

24. F. TODA, H. MIYAMOTO and S. KIKUCHI. *J. Chem. Soc., Chem. Commun.* (1995) 621.

25. K. TANAKA and F. TODA. *J. Chem. Soc., Perkin Trans.* **1** (1992) 943.

26. J. N. MOORTHY and K. VENKATESAN. *J. Org. Chem.* **56** (1991) 6957.

27. F. TODA, K. TANAKA and M. ODA. *Tetrahedron Lett.* **29** (1988) 653.

28. T. FUJIWARA, N. NANBA, K. HAMADA, F. TODA and K. TANAKA. *J. Org. Chem.* **55** (1990) 4532.

29. F. TODA, H. MIYAMOTO and K. KANEMOTO. *J. Org. Chem.* **61** (1996) 6490.

30. F. TODA, H. MIYAMOTO, S. KIKUCHI, R. KURODA and F. NAGAMI. *J. Am. Chem. Soc.* **118** (1996) 11315.

31. F. TODA, H. MIYAMOTO and K. KANEMOTO. *J. Chem. Soc., Chem. Commun.* (1995) 1719.; F. TODA and H. MIYAMOTO. *Chem. Lett.* (1995) 809.

32. F. TODA, H. MIYAMOTO, H. KOSHIMA and Z. URBANCZYK-LIPKOWSKA. *J. Org. Chem.* **62** (1997) 9261.

33. K. TANAKA and F. TODA. *Mol. Cryst. Liq. Cryst.* **313** (1998) 179.

34. K. TANAKA, F. TODA, E. MOCHIZUKI, N. YASUI, Y. KAI, I. MIYAHARA and K. HIROTSU. *Angew. Chem. Int. Ed. Engl.* **38** (1999) 3523; K. TANAKA, E. MOCHIZUKI, N. YASUI, Y. KAI, I. MIYAHARA, K. HIROTSU and F. TODA. *Tetrahedron* **56** (2000) 6853.

35. K. TANAKA, O. KAKINOKI and F. TODA. *J. Chem. Soc., Perkin Trans.* **1** (1992) 307.

36. K. TANAKA, H. MIZUTANI, I. MIYAHARA, K. HIROTSU and F. TODA. *J. Chem. Soc., Cryst. Eng. Comm.* (1999) 3.

37. F. TODA, M. YAGI and S. SODA. *J. Chem. Soc., Chem Commun.* (1987) 1413; A. SEKINE, K. HORI, Y. OHASHI, M. YAGI and F. TODA. *J. Am. Chem. Soc.* **111** (1989) 697.

38. F. TODA and H. MIYAMOTO. *J. Chem. Soc, Perkin Trans.* **1** (1993) 1129; D. HASHIZUME, H. KOGO, A. SEKINE, Y. OHASHI, H. MIYAMOTO and F. TODA. *J. Chem. Soc., Perkin Trans.* **2** (1996) 61; D. HASHIZUME, H. KOGO, A. SEKINE, Y. OHASHI, H. MIYAMOTO and F. TODA. *Acta Cryst.* **C51** (1995) 929; T. ASAHI, M. NAKAMURA, J. KOBAYASHI, F. TODA and H. MIYAMOTO. *J. Am. Chem. Soc.* **119** (1997) 3665; D. HASHIZUME, H. KOGO, Y. OHASHI, H. MIYAMOTO and F. TODA. *Anal. Sci.* **14** (1998) 1187.

39. F. TODA and K. TANAKA. *Supramol. Chem.* **3** (1994) 87; F. TODA, K. TANAKA, Z. STEIN and I. GOLDBERG. *Acta Cryst.* **B51** (1995) 856.; F. TODA, K. TANAKA, Z. STEIN and I. GOLDBERG. *Acta Cryst.* **C51** (1995) 2722.

Achieving Enantio and Diastereoselectivities in Photoreactions Through the Use of a Confined Space

J. Sivaguru, J. Shailaja, S. Uppili, K. Ponchot, A. Joy, N. Arunkumar and V. Ramamurthy

Department of Chemistry, Tulane University, New Orleans, LA 70118, USA

1. Introduction

The efforts of chemists during the past few decades have advanced the field of thermal asymmetric synthesis to a great extent [1]. Complex molecules can now be synthesized as single enantiomers. Unfortunately, asymmetric photochemical reactions have not enjoyed the same level of success [2]. In the past, chiral solvents, chiral auxiliaries, circularly polarized light, and chiral sensitizers have been utilized to conduct enantioselective photoreactions. The highest chiral induction achieved by any of these approaches at ambient temperature and pressure has been ~30% (2–10% e.e. is common in photochemical reactions under the above conditions). Crystalline state and solid host-guest assemblies have, on the other hand, provided the most encouraging results [3]. Two approaches have been used to achieve chiral induction in the crystalline state. In one, by the Weizmann Institute Group, the achiral reactant is crystallized into a chiral space group [4]. The limited chances of such crystallization of organic molecules renders this approach less general. In the second approach, due to Scheffer and co-workers [5], an ionic chiral auxiliary is used to effect a chiral environment. This limits the approach to molecules with carboxylic acid groups that form crystallizable salts with chiral amines or *vice versa*. Yet another successful approach due to Toda and co-workers [6] has made use of organic hosts that contain chiral centers (e.g., deoxycholic acid, cyclodextrin, 1,6-bis (o-chlorophenyl)-1,6-diphenyl-2,4-diyne-1,6-diol,). The success of this approach is limited to guests that can form solid solutions with the host without disturbing the hosts macro-structure. The reactivity of molecules in the crystalline state and in solid host-guest assemblies is controlled by the details of molecular packing. Currently, molecular packing and consequently the chemical reactivity in the crystalline state, can not be reliably predicted [7]. Therefore even after successfully crystallizing a molecule in a chiral space group or complexing a molecule with a chiral host or a chiral auxiliary, there is no guarantee that the guest will react in the crystalline state. Hence, even though crystalline and host-guest assemblies have been very useful in conducting enantioselective photoreactions, their general applicability thus far has been limited.

Zeolites, we believe, offer a solution to the above limitations [8]. For example, guest molecules present within zeolites possess greater freedom than in crystals/solid host-guest assemblies and less freedom than in isotropic solvents. This permits molecules within zeolites to react more freely than in crystals/solid host-guest assemblies and more selectively than in solutions. Zeolites, unlike organic host systems can include a large number of molecules, the only limitation being that the dimensions of the guest be less than the pore dimensions of the zeolite. A variety of molecules can therefore be

F. Toda (ed.), Organic Solid-State Reactions, 159–188.

included within a zeolite and induced to react. A major deficiency is that the zeolite is not chiral [9]. In order to be a useful chiral medium this serious limitation has to be overcome and our approach in this context has been to chirally modify the zeolite. We have adopted several strategies to achieve chiral induction on photoreactions that yield racemic products in solution. These are briefly summarized in this chapter.

2. Zeolite

Zeolites are crystalline aluminosilicates with open framework structures made up of the primary building blocks $[SiO_4]^{4-}$ and $[AlO_4]^{5-}$ tetrahedra that are linked by all their corners to form channels and cages of discrete size with no two aluminum atoms sharing the same oxygen. Due to the difference in charge between the $[SiO_4]^{4-}$ and $[AlO_4]^{5-}$ tetrahedra, the net framework charge of an aluminum-containing zeolite is negative and hence must be balanced by a cation, typically an alkali or alkaline earth metal cation. The adsorbed molecules are located in the cavities of the zeolites and access to these cavities is through a pore or window whose size can be the same or smaller than the cavities. The dimension of this pore determines the size of the molecule that can be adsorbed into these structures. We have utilized X and Y zeolites as media for our chiral studies and the structure of these are mentioned below. Faujasite zeolites X and Y are large pore zeolites with the following unit cell composition:

$$X \text{ type } M_{86}(AlO_2)_{86}(SiO_2)_{106}.264 \text{ H}_2O$$
$$Y \text{ type } M_{56}(AlO_2)_{56}(SiO_2)_{136}.253 \text{ H}_2O$$

where M is the charge-compensating cation. These cations occupy three different positions within the zeolites X and Y (Figure 1). Of the three types of cation locations only types II and III are located within the supercage where an organic molecule is expected to reside. Due to the hygroscopic nature of the cations zeolites tend to adsorb considerable

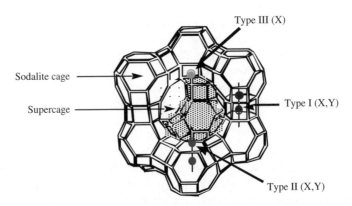

Figure 1 The basic structural unit (supercage) of X and Y zeolites. Cation locations within a supercage are shown as dark spheres. The supercage is connected to four adjacent supercages through the ~7.4 å windows (one window is facing the reader).

amounts of water (see the formulae). These water molecules can be desorbed through activation at high temperatures creating space for organic guest molecules to be included. The free volume available for the adsorbed molecule within the supercage depends on the number and nature of the cation (for example, it varies from $873 Å^3$ in NaX to $732 Å^3$ in CsX).

2.1. A chiral zeolite – local versus global chirality

An ideal approach to achieving chiral induction in a constrained medium such as zeolite would be to make use of a chiral medium. To our knowledge no zeolite that can accommodate organic molecules, currently exists in a stable chiral form [9]. Though zeolite beta and titanosilicate ETS-10 have unstable chiral polymorphs, no pure enantiomorphous forms have been isolated. Although many other zeolites can, theoretically, exist in chiral forms (e.g., ZSM-5 and ZSM-11) none has been isolated in such a state. In the absence of readily available chiral zeolites, we are left with the choice of creating an asymmetric environment within zeolites by the adsorption of chiral organic molecules. We have achieved this through two approaches: (a) by including an optically pure organic molecule as an adsorbent within a zeolite and (b) exchanging the inorganic cations present in zeolites with chiral ammonium ions.

One might question the long-term usefulness of the strategy of chirally modifying an achiral zeolite (as opposed to synthesizing a chiral zeolite). One is also likely to argue that possible future availability of chiral zeolites will make this approach obsolete. In this context it is important to recognize that solvents, liquid crystals and micelles that are optically pure are yet to yield respectable chiral induction when used as media for photoreactions suggesting that the 'global' chirality of the medium alone may not be sufficient to effect chiral induction of a reaction. It is our opinion that in order to achieve chiral induction, 'local' chirality is more important than 'global' chirality and the strategy of functionalizing a zeolite with a chiral inductor will result in such a local chirality.

The chiral inductor that is used to modify the zeolite interior will determine the magnitude of the enantioselectivity of the photoproduct. The suitability of a chiral inductor for a particular study depends on its inertness under the given photochemical condition, its shape, size (in relation to that of the reactant molecule and the free volume of the zeolite cavity) and the nature of the interaction(s) that will develop between the chiral agent and the reactant molecule/transition state/reactive intermediate. One should recognize that no single chiral agent might be ideal for two different reactions or at times structurally differing substrates undergoing the same reaction. These are inherent problems of chiral chemistry. Lack of sufficient examples in the literature necessitates a certain amount of empiricism at this stage. Thus pursuit of a path of 'rational combinatorial approach' until ground rules are established is a wise choice.

3. A Noncovalent 'Chiral Inductor' Strategy

The strategy of employing chirally modified zeolites as reaction media requires inclusion of two different molecules, the chiral inductor and the achiral reactant within the interior

161

spaces of the achiral zeolite. When two different molecules are included within a
zeolite, the distributions are expected to follow the pattern shown in Figure 2.

The six possible distributions of guest molecules are represented by cages labeled
A to F. A and C can give only racemic mixture of products. B can induce chiral induction
to a certain extent depending on the interaction between the chiral inductor and the achiral molecule. The rest of the cages will not participate in the desired reaction as they do
not contain the achiral reactant. Hence the total chiral induction is a sum of A, B and C.

The chiral inductors used in these investigations are selected based on several criteria,
including availability, size and functionality. An inductor has to be small enough to fit
into the interior of the zeolite and small enough to accommodate another guest molecule
in the same supercage. Also, a chiral inductor should have functional groups that are
capable of interacting with the zeolitic interior, as well as with the functional groups of
the substrate. Figure 3 lists the chiral inductors explored.

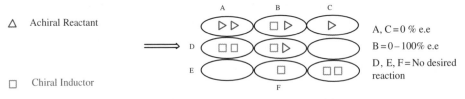

Figure 2 Possible distributions of substrate and chiral inductor inside zeolite.

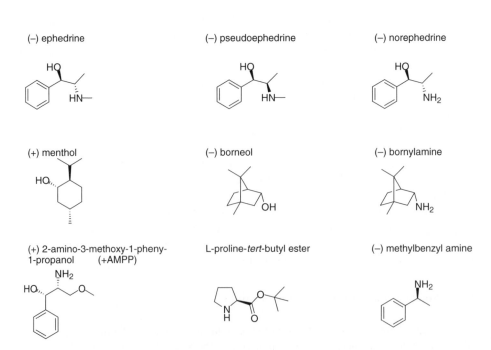

Figure 3 Continued.

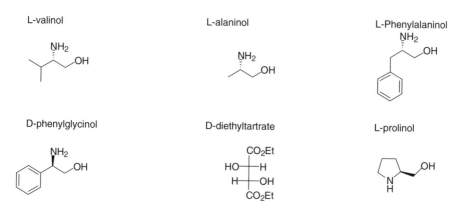

L-valinol

L-alaninol

L-Phenylalaninol

D-phenylglycinol

D-diethyltartrate

L-prolinol

Figure 3   Chiral inductors used to chirally modify zeolites.

3.1.   Chiral induction during Norrish–Yang photochemical reactions

In the context of chiral induction in photochemical reactions, we have examined the Norrish–Yang reaction, which has been extensively studied in isotropic and anisotropic media [10]. The mechanism of the reaction consists of intramolecular γ-hydrogen abstraction by an excited carbonyl oxygen atom to produce a triplet 1,4-biradical that has three possible reaction pathways: closure to give cyclobutanol, cleavage to an alkene and enol (isolated as a corresponding keto compound), and reverse hydrogen transfer to regenerate the ground-state ketone (**Scheme 1**). Both *cis*- and *trans*- cyclobutanols exist

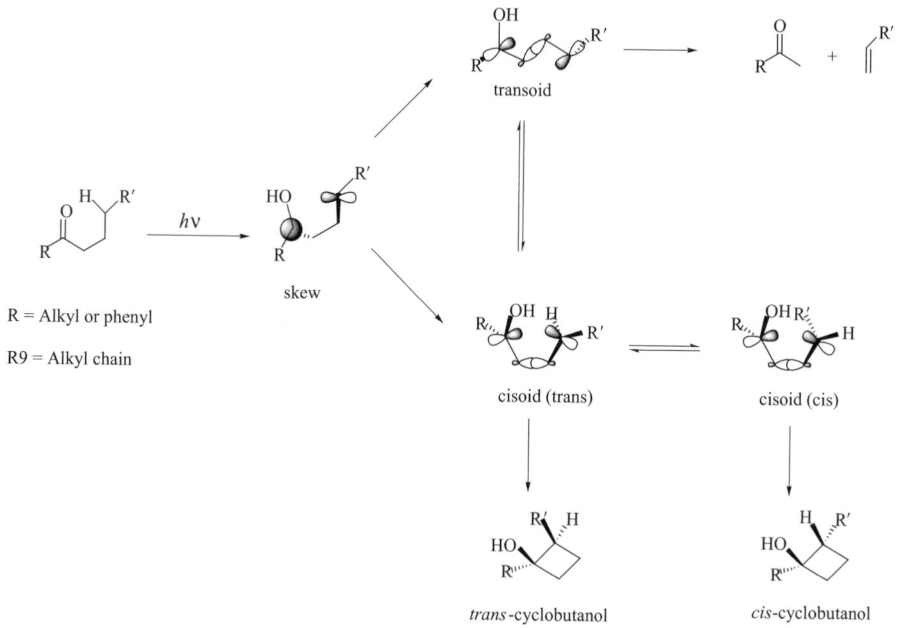

R = Alkyl or phenyl

R9 = Alkyl chain

transoid

skew

cisoid (trans)

cisoid (cis)

trans-cyclobutanol

cis-cyclobutanol

Scheme 1

as pairs of enantiomers. These reactions are good candidates for photoreactions inside zeolites, wherein under appropriate conditions, one of the enantiomeric cyclobutanols can be formed preferentially. The results of asymmetric Norrish–Yang reaction of several cyclohexyl and adamantyl aryl ketones are summarized below.

Scheme 2 represents the conversion of *cis*-4-*tert*-butylcyclohexyl ketones of general structure **1** into the corresponding cyclobutanols of structure **2** [11]. Ephedrine proved to be the best chiral inductor giving 25–30% enantiomeric excess. The use of (+)-ephedrine afforded the optical antipode of the photoproduct produced by the use of (−)-ephedrine, indicating that the system is well behaved. Results of photolysis with a variety of chiral inductors are listed in Table 1.

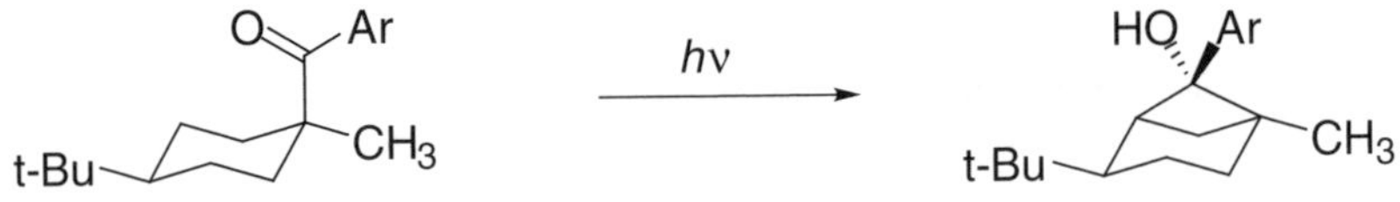

(a) Ar=*p*-Ph-CN
(b) Ar=*p*-Ph-COOMe

Scheme 2

By employing the chiral inductor approach, modest enantioselectivity has been achieved. A tight fit between the reactant and the chiral inductor may be a prerequisite to achieve significant enantioselectivity. While in MY zeolite, (−)ephedrine seems to show enantioselectivity in favor of the (+)-isomer, the same chiral inductor favors, although to much less extent, the (−)-isomer in NaX (**Scheme 3**). We attribute this effect to the difference in free volume between NaX and NaY and the number of type III cations. We speculate that the larger dimension of the supercages in NaX (852 Å^3) favors an orientation between the reactant and the chiral inductor that is different from NaY. Supporting this interpretation are the results obtained with KX (800 Å^3), RbX (770 Å^3) and CsX (732 Å^3). As the cation gets larger, the free volume inside the supercage decreases. When the supercage volume was reduced until it was the same or smaller than that of NaY the enantioselectivity was found to be the same as NaY.

Table 1. Asymmetric induction in the Norrish–Yang reaction of aryl ketones

Ketone	Chiral inductor	% e.e.[a]
1a	(−)-ephedrine	25B
1b	(−)-ephedrine	30B
1b	(−)-menthol	8A
1b	(−)-borneol	9A
1b	(+)-bornylamine	6A
1b	L-proline *tert*-butyl ester	<5B

[a] The first peak to elute from the column is assigned as A and the second as B.

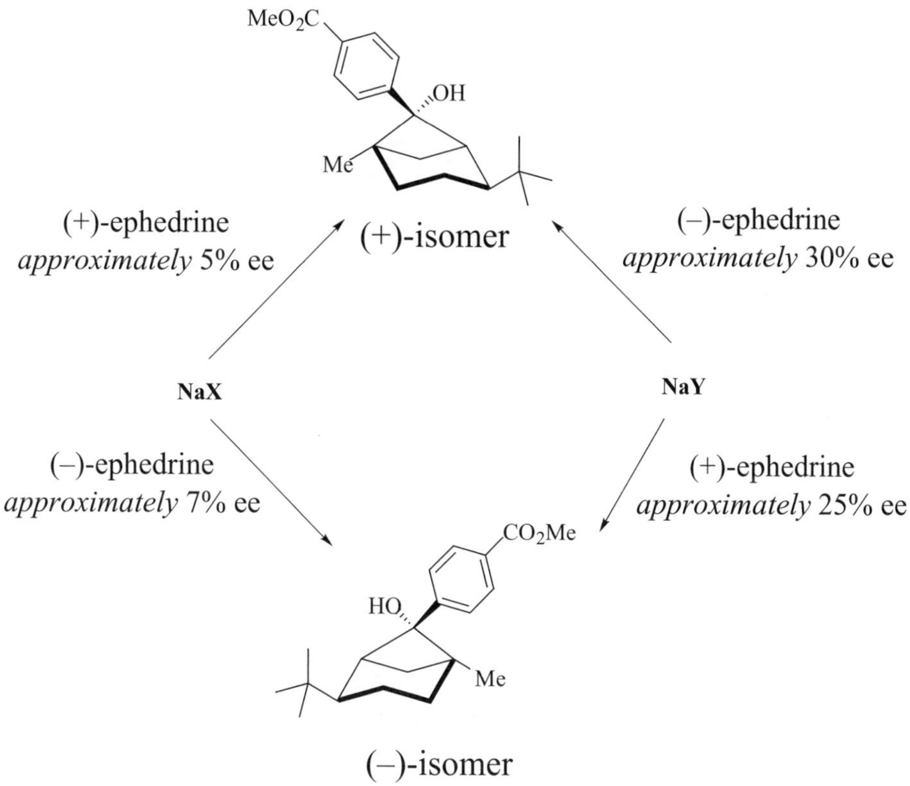

Scheme 3

Ketone **3**, shown below, undergoes Norrish–Yang reaction giving cyclobutanols as products (**Scheme 4**) [12]. Unlike **1b**, which gave only one cyclobutanol diastereomer both in solution as well as in zeolite, ketone **3** gave both *cis*-cyclobutanol, **4**, and *trans*-cyclobutanol, **5**. With NaY loaded with one (−)-ephedrine molecule per supercage, the ratio of **5** to **4** is 1.4. While **4** is formed in moderate enantioselectivity (35%), **5** was formed in low enantioselectivity (5%).

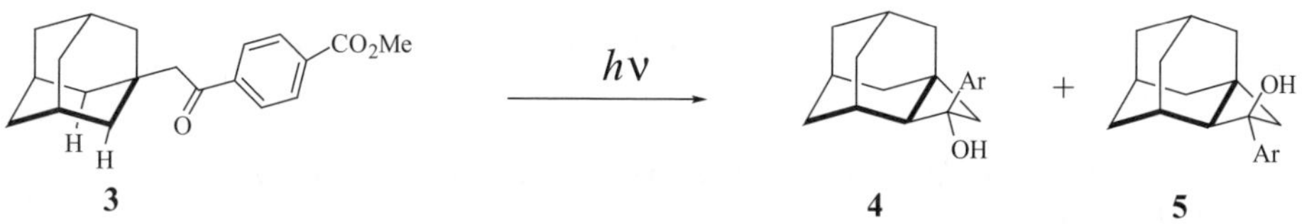

Scheme 4

The irradiation of phenyl-2-methyltricyclo[3.3.1.1]dec-2-yl methanone **6a** (**Scheme 5**) was carried in faujasite in the presence of chiral inductors [13]. The maximum enantio-selectivity obtained using the chiral inductor approach was 32% ee both with (−)-pseudoephedrine and (+)-2-amino-3-methoxy-1-phenyl-1-propanol. Similar studies were

165

Table 2   Chiral induction with various inductors during the irradiation of adamantyl phenyl ketones, **6a** and **6b**, in NaY

Chiral inductor	% ee[a]	Chiral inductor	% ee[a]
		Adamantyl phenyl ketone, **6a**	
(–)-pseudephedrine	32 B	(+) pseudoephedrine	28 A
(–)-norephedrine	15 B	(+)-norephedrine	12 A
(–)-ephedrine	2 A	(+) AMPP	32 A
(+)-methylbenzylamine	22 A	L-valinol	7 B
L-phenylalaninol	14 B	L-alaninol	3 B
D-phenylglycinol	racemic	L-proline-*tert*-butyl ester	5 A
(+)-diethyltartrate	11 B		
		Adamantyl phenyl ketone, **6b**	
(–)-pseudoephedrine	30 B	(+)-pseudoephedrine	20 A
(–)-norephedrine	2 A	(–)-ephedrine	racemic
D-phenylglycinol	4 B	L-phenylalaninol	5 B
(+)-methylbenzylamine	3 A		

[a] The first peak to elute from the column is assigned as A and the second as B.

done for 1-(4-fluorophenyl)-2-methyltricyclo[3.3.1.1]dec-2-yl methanone **6b**. The maximum enantioselectivity obtained in this case was 30% ee with (–)-pseudoephedrine. The results that were obtained using various chiral inductors are tabulated in Table 2.

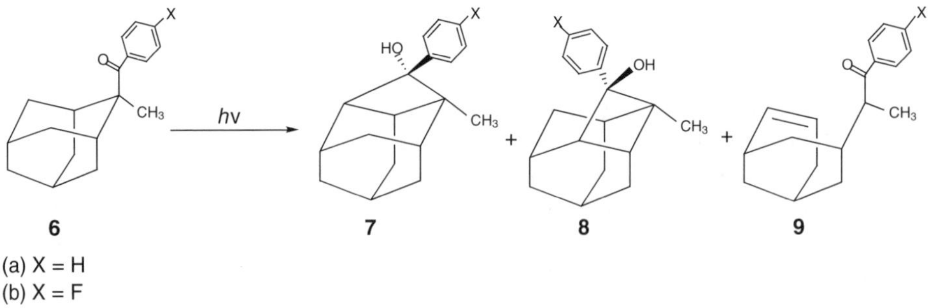

6
(a) X = H
(b) X = F

7   **8**   **9**

Scheme 5

3.2.   Chiral induction during the photoisomerization of diphenylcyclopropane derivatives

Ethyl ester **10** undergoes photoisomerization as shown in **Scheme 6** [14]. The reaction in isotropic medium gave a racemic mixture of the *trans*-isomer **11** and **12**. Enantiomeric excess of 17% was achieved by the chiral inductor approach with cyclohexyl ethyl amine [15]. The results with various chiral inductors are summarized in Table 3.

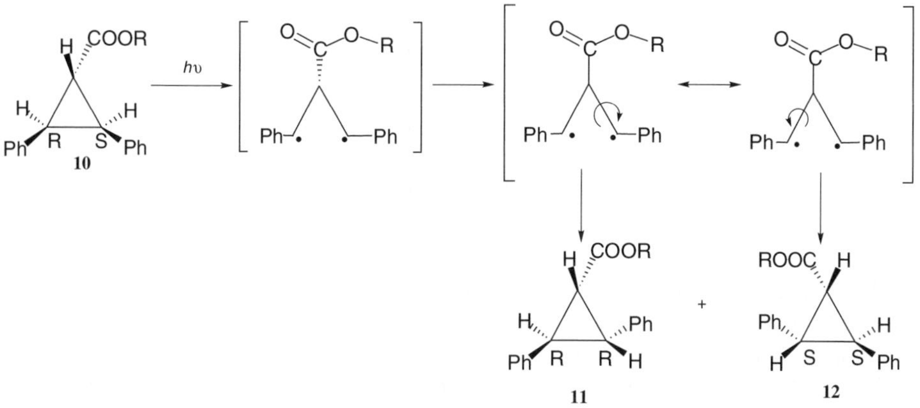

Scheme 6

Compound **13** is similar to **10**, except that the ester group is changed to a keto group. Upon, excitation, **13** will be converted to the chiral *trans*-isomers **14** and **15** as shown **Scheme 7** [16].

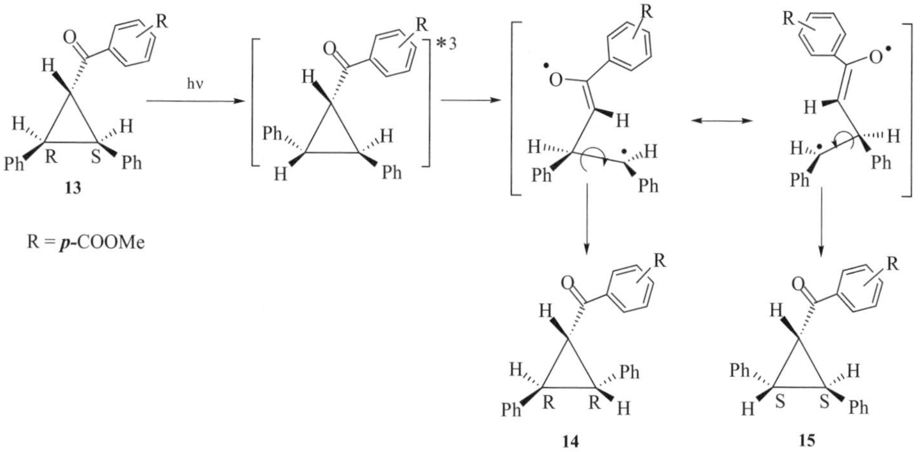

R = *p*-COOMe

Scheme 7

Table 3 Chiral induction during the *cis*- to *trans*-photoisomerization of **10** in NaY

Chiral inductor	% e.e.[a]	Chiral inductor	% e.e.[a]
(+)-pseudoephedrine	5B	(−)-cyclohexylethyl amine	17A
R(+)-bornylamine	4A	(−)-diethyltartrate	12B
(−)-menthol	racemic	(+)-norephedrine	5B
(−) ephedrine	racemic	L-phenylalaninol	5B
D-valinol	racemic	L-alaninol	4A

[a] The first peak to elute from the column is assigned as A and the second as B.

Table 4 Chiral induction during the *cis*- to *trans*-photoisomerization of **13** in NaY

Chiral inductor	% ee[a]	Chiral inductor	% ee[a]
(–)-norephedrine	20 A	(–)-diethyltartrate	3 B
(+)-norephedrine	20 B	L-phenylalaninol	7 A
(+)-pseudoephedrine	8A	(–)-ephedrine	3 A
(–)-pseudoephedrine	10 B	(+)-menthol	2 B

[a] The first peak to elute from the column is assigned as A and the second as B.

Chiral induction for compound **13** with various chiral inductors inside NaY zeolite is given in Table 4. 20% enantiomeric excess was achieved with norephedrine as the chiral inductor. In both the cases (compound **10** and **13**), when the optical antipode of the chiral inductor was used, the opposite enantiomer was in excess. In both the cases the enantioselectivity was low [15].

3.3. Chiral induction during the di-π-methane rearrangement of cyclohexadienones and dihydronaphthalenone systems

Photochemistry of substituted cyclohexadienone, **16**, is dependent on the polarity of the reaction medium (**Scheme 8**) [17]. For instance, in nonpolar solvents like hexane the reaction occurs from the nπ* state that results in a ketene, which can be trapped by ethanol or diethyl amine. In highly polar solvents like trifluoroethanol, the reaction occurs from the $\pi\pi$* excited state and results in an oxa-di-π-methane rearranged product. Cations present within the zeolite cavities ensure the formation of the desired $\pi\pi$* product that exists in two enantiomeric forms, **17** and **18**. It was found that the enantiomeric excess was dependent on the chiral inductor (Table 5), the water content, and the cation present in these zeolites (Table 6) [18]. Also, an unprecedented temperature

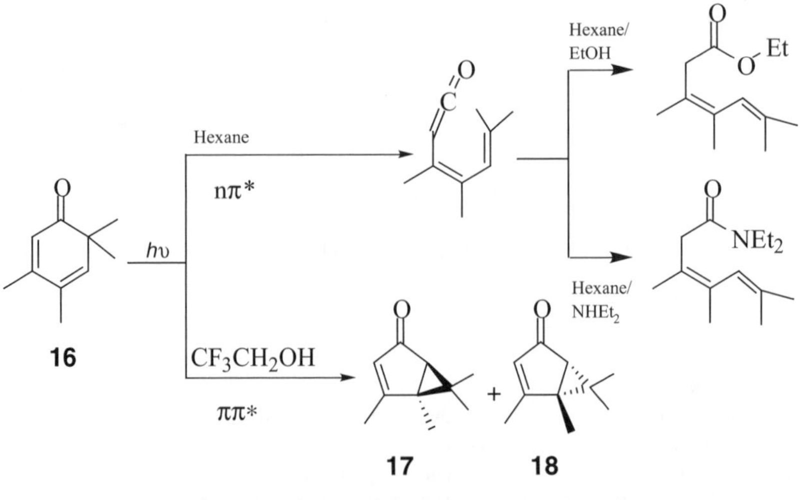

Scheme 8

Table 5 Chiral induction during the photochemistry of **16** in NaY

Chiral inductor	% ee[a]	Chiral inductor	% ee[a]
none	racemic	(+)-methylbenzylamine	racemic
(−)-ephedrine	30 B	(−)-methylbenzylamine	racemic
(+)-ephedrine	28 A	(−)-menthol	10 B
(−)-pseudoephedrine	28 B	(+)-menthol	8 A
(+)-pseudoephedrine	20 A	(−)-diethyltartrate	9 A
(−)-norephedrine	14 A	(+)-diethyltartrate	12 B
(+)-norephedrine	17 B		

[a] The first peak to elute from the column is assigned as A and the second as B.

Table 6 Effect of cation on chiral induction of **16**. Inductor: (−)-ephedrine

Cation in Y-zeolite	% ee[a]
Li^+	racemic
Na^+	32 B
K^+	26 B
Rb^+	5 A
Cs^+	racemic

[a] The first peak to elute from the column is assigned as A and the second as B.

Table 7 Effect of temperature in chiral induction of **16**

NaY/(−)-ephedrine		NaY/(+)-ephedrine	
Temperature/°C	% ee[a]	Temperature/°C	% ee[a]
88	7 B	96	3 A
25	30 B	25	25 A
0	35 B	−7	32 A
−55	49 B	−48	40 A

[a] The first peak to elute from the column is assigned as A and the second as B.

dependence of the enantioselectivity was observed for this system (Table 7). The % ee increases with decrease in temperature.

1,2-Dihydro-2,2-dimethylnaphthalenone systems, **19**, have a similar chromophore as that of **16**. The achiral 1,2-dihydro-2,2-dimethylnaphthalenone **19a** undergoes di-π-methane rearrangement to chiral benzobicyclo[3.1.0]hexenones **20** and **21** (**Scheme 9**) [19]. This system yields a racemic mixture of the photoproducts when irradiated in acetonitrile both in the presence and absence of chiral inductors. The % ee is dependent on the chiral inductor (Table 8), the water content within the supercage and the cations present in the zeolite. The highest % ee obtained with these systems was 17% in NaY with (+)-pseudoephedrine as the chiral inductor. With the methyl ester derivative **19b**, the observations were similar to that of **19a** and the maximum % ee obtained was 18% with NaY and (–)-ephedrine (Table 9) [20]. Also the % ee was dependent on the type of cation present in the zeolite (Table 10). Unlike cyclohexadienone systems, the enantiomeric excess in the napthalenones was independent of the reaction temperature (Table 11).

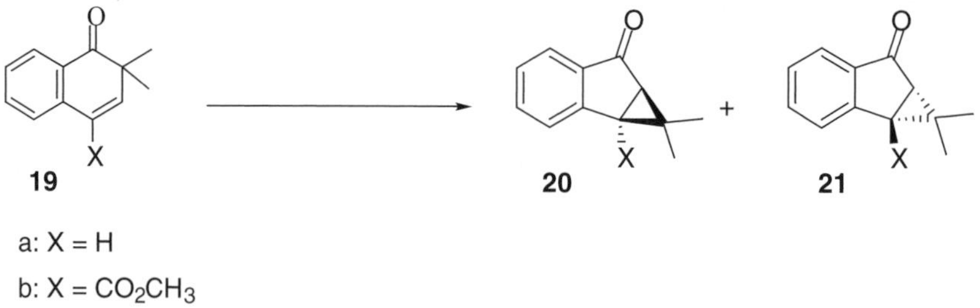

19

a: X = H

b: X = CO$_2$CH$_3$

Scheme 9

Table 8   Chiral induction during the photochemistry of **19a** in NaY

Chiral inductor	% ee[a]	Chiral inductor	% ee[a]
(–)-ephedrine	13 A	(+)-ephedrine	10 B
(–)-norephedrine	3 A	(+)-norephedrine	2 B
(–)-pseudoephedrine	15 B	(+)-pseudoephedrine	17 A
(–)-menthol	13 A	(+)-menthol	12 B
L-phenylalaninol	2 B	D-phenylglycinol	15 B
none	racemic		

[a] The first peak to elute from the column is assigned as A and the second as B.

Table 9   Chiral induction during the photochemistry of **19b** in NaY

Chiral inductor	% ee[a]	Chiral inductor	% ee[a]
(+)-ephedrine	15 A	(–)-ephedrine	18 B
(+)-pseudoephedrine	17 B	(–)-pseudoephedrine	14 A
(+)-methylbenzylamine	5 A	(–)-methylbenzylamine	racemic
L-phenylalaninol	2 B	(+)-pseudoephedrine•HCl	12 B
none	racemic		

[a] The first peak to elute from the column is assigned as A and the second as B.

Table 10 Effect of cation on enantioselectivity during photochemistry of **19b** in M⁺Y/(−) ephedrine

Table 10 Effect of cation on enantioselectivity during photochemistry of **19b** in $M^+Y/(-)$ ephedrine

Cation in Y-zeolite	% ee[a]
Li⁺	18 B
Na⁺	17 B
K⁺	10 A
Rb⁺	8 A
Cs⁺	11 A

[a] The first peak to elute from the column is assigned as A and the second as B.

Table 11 Effect of temperature on enantioselectivity during photochemistry of **19b** in NaY/(−)-ephedrine

Temperature/°C	% ee[a]
0	13 B
−20	14 B
−40	10 B

[a] The first peak to elute from the column is assigned as A and the second as B.

3.4. Cyclization of tropolone ethers

Upon excitation tropolone alkyl ethers undergo 4π electrons disrotatory ring closure to yield racemic products; depending on the mode of disrotation opposite enantiomers are obtained (**Scheme 10**) [21,22]. Enantiomerically pure products can be obtained by controlling the mode of ring closure. Based on this rationale we examined the photochemistry of tropolone alkyl ethers within chirally modified zeolites. The enantiomeric excesses (e.e.) obtained with tropolone alkyl ethers are dependent on the chiral agent (**Scheme 11**), the alkoxy substituent (**Scheme 12**), the water content within the supercage (**Scheme 13**), the nature and number of cations (**Scheme 13**), and the temperature [23]. The results obtained with tropolone alkyl ethers clearly illustrate the complexities of chiral induction within a zeolite and point out the parameters that need to be optimized to obtain high e.e. The high e.e. (69%) obtained with tropolone ethyl phenyl ether is most encouraging (**Scheme 13**).

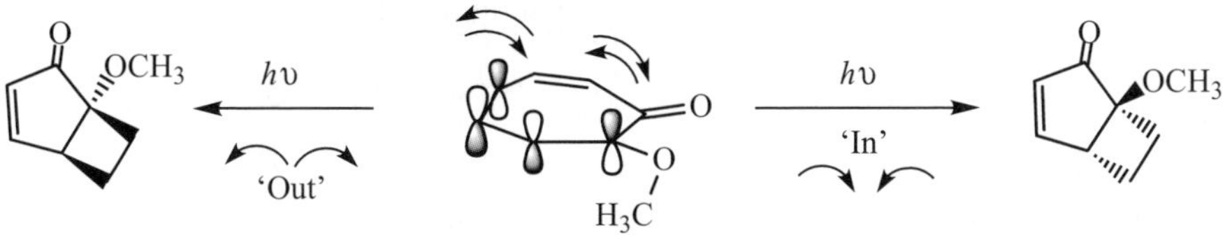

Scheme 10

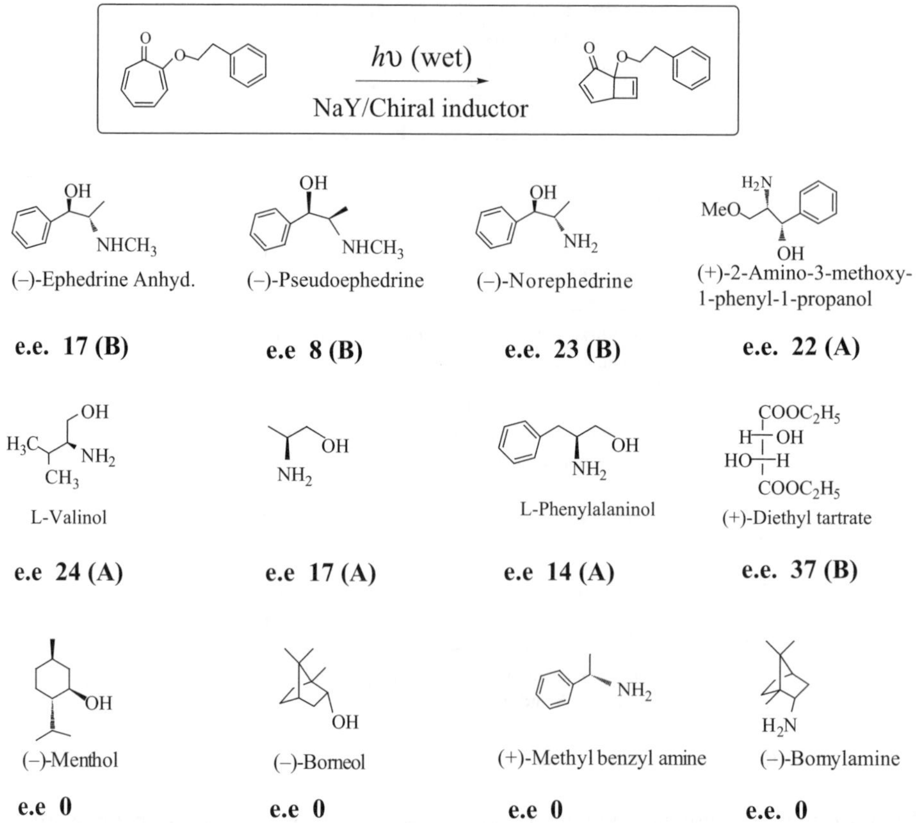

Scheme 11

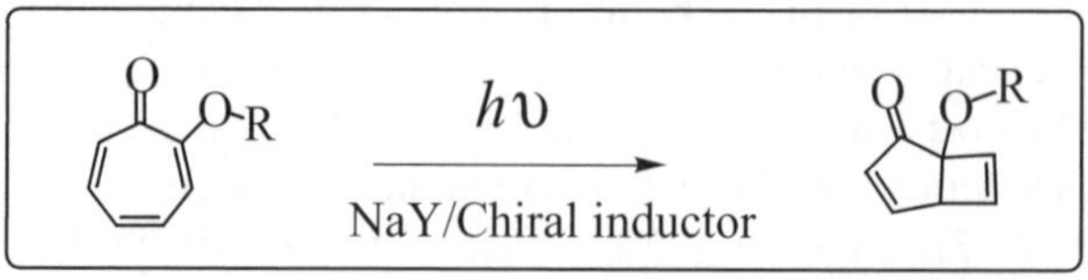

	Chiral Inductor	ee %	Chiral Inductor	ee %
R = CH$_3$	Ephedrine	17%	Norephedrine	35%
R = CH$_2$CH$_3$	Ephedrine	8%	Norephedrine	11%
R = CH(CH$_3$)$_2$	Ephedrine	0%	Norephedrine	6%

Scheme 12

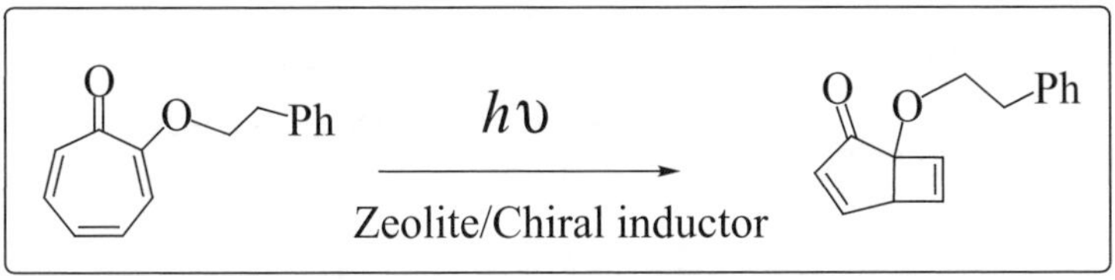

Chiral inductor	Enantiomeric excess (%)	
	'wet'	'dry'
Na Y/ (+)Ephedrine	17 (A)	69 (B)
Na Y/(–) Pseudoephedrine	8 (B)	20 (A)
Na Y/(–) Norephedrine	23 (B)	38 (A)
Na X/ (+)Ephedrine	9 (B)	
Li Y/ (+)Ephedrine		22 (B)
KY/ (+)Ephedrine		11 (B)
Rb Y/ (+)Ephedrine		2 (B)

Scheme 13

3.5. Miscellaneous reactions

In addition to the above reactions we have examined the Schenk-ene reaction (**Scheme 14**) and the Zimmerman di-π-methane rearrangement (**Scheme 15**) within zeolites, and the choice has been dictated by the ease of synthesis of reactants, the feasibility of separation of chiral products and the amenability of the reaction for spectroscopic investigations [24]. The photoreactions *per se* are not of paramount importance, but the information they provide will help to establish the concept and generate a model for future experiments.

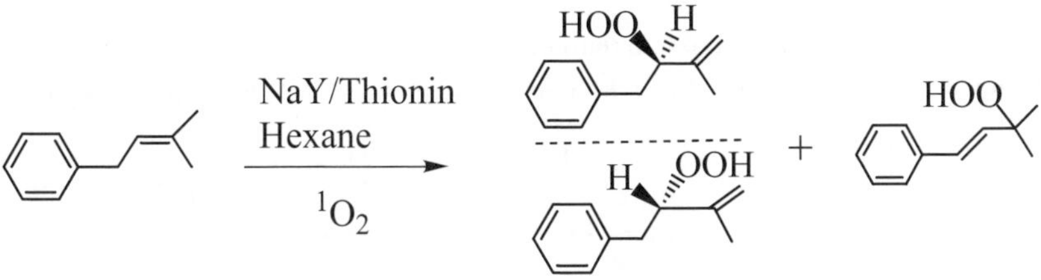

No chiral inductor 0–0.05% ee
Ephedrine hydrochloride 15.5% ee

Scheme 14

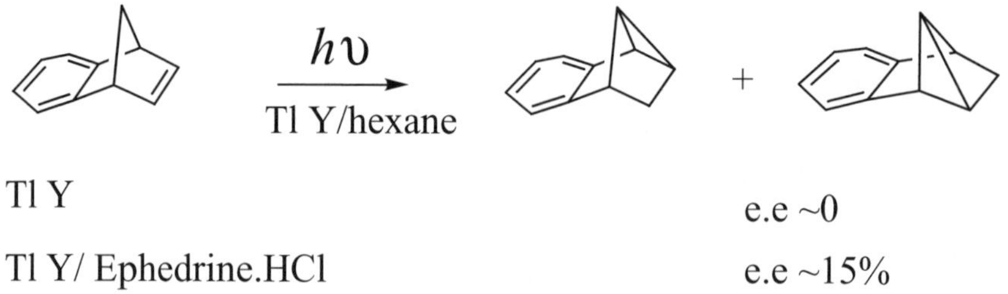

$$h\upsilon \xrightarrow{\quad} \quad + $$

Tl Y/hexane

Tl Y   e.e ~0

Tl Y/ Ephedrine.HCl   e.e ~15%

Scheme 15

4. Covalent 'Chiral Auxiliary' Strategy

One of the main drawbacks of the chiral inductor strategy discussed above is that the extent of the chiral induction depends on the statistical distribution of the reactants and chiral inductor. To achieve high enantiomeric excess, both the chiral inductor and the reactant have to be near one another within a zeolite. We reasoned that unless a strategy to place every reactant molecule next to a chiral inductor within a zeolite is developed, high stereoselectivity is unlikely by the above chiral inductor approach. This led us to explore the chiral auxiliary approach.

In this approach, the chiral inductor is connected to the reactant via a covalent bond. This forces the chiral inductor and reactant parts to be in close proximity most of the time. Even these molecules can adopt different conformations. The distribution of the chirally modified reactant molecule is shown in Figure 4. There can be two possible distributions. In the first case, the entire molecule is in the same supercage of the zeolite, which will lead to between 0–100%de depending on the extent of interaction between the chiral auxiliary and the reaction center (Type I). In the second case, the chiral auxiliary and the reactant may be stretched between the two cages (Type II). Such situations are most likely to yield 1:1 diastereomeric mixture of products. Auxiliaries used were chiral alcohols or amines that were covalently attached to substrates as ester or amide linkages. Figure 5 lists the chiral auxiliaries explored. We have tested this approach with various systems and the results are briefly presented here.

4.1. Chiral auxiliary approach with adamantyl phenyl ketones

The Norrish–Yang reaction of 2-benzoyladamantane-2-carboxylic acid derivative **22** is shown in **Scheme 16** [25]. Chiral auxiliaries were introduced as ester functionalities.

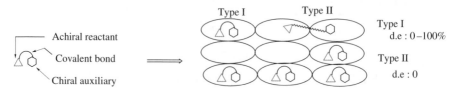

Figure 4 Possible distributions of substrates with chiral auxiliaries in zeolite.

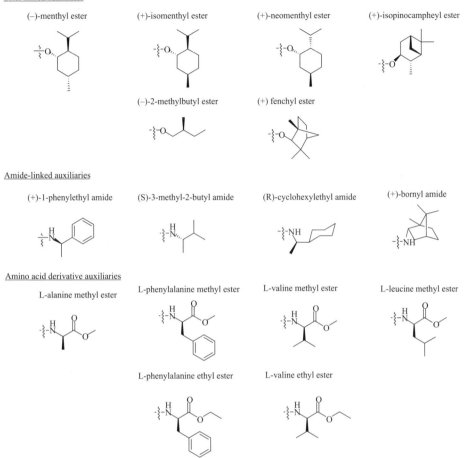

Ester linked auxiliaries

(–)-menthyl ester (+)-isomenthyl ester (+)-neomenthyl ester (+)-isopinocampheyl ester

(–)-2-methylbutyl ester (+) fenchyl ester

Amide-linked auxiliaries

(+)-1-phenylethyl amide (S)-3-methyl-2-butyl amide (R)-cyclohexylethyl amide (+)-bornyl amide

Amino acid derivative auxiliaries

L-alanine methyl ester L-phenylalanine methyl ester L-valine methyl ester L-leucine methyl ester

L-phenylalanine ethyl ester L-valine ethyl ester

Figure 5   Chiral auxiliaries employed.

Irradiation of these esters gave **24** as final products via cyclobutanol **23**. Irradiation of these ketoesters in acetonitrile gave 5–14%de for different auxiliaries. When the same molecules were irradiated in MY zeolite, up to 79%de was achieved in LiY for **22a**, and NaY for **22c**. The extent of %de varied with cation (Table 12).

22

(a) menthyl ester
(b) 2-methylbutyl ester
(c) isomenthyl ester
(d) isopinocampheyl ester

23   **24**

Scheme 16

175

Table 12   Diastereoselectivity during Norrish–Yang reaction of **22a–d**

Substrate	%de[a] in various media			
	solution	LiY	NaY	KY
(–)-menthyl ester, **22a**	5 B	79 B	60 B	31 B
(–)-2-methylbutyl ester, **22b**	1 B	26 B	54 B	30 B
(+)-isomenthyl ester, **22c**	14 B	65 A	79 A	3 A
(+)-isopinocampheyl ester, **22d**	5 B	52 B	62 B	4 A

[a] The first peak to elute from the column is assigned as A and the second as B.

4.2. Photoisomerization of diphenylcyclopropane derivatives

Compound **10** (from **Scheme 6**) was chirally modified by replacing its ethyl ester with chiral auxiliaries to give various esters **25**, amides **26**, and amino acid derivatives **27** (**Scheme 17**).

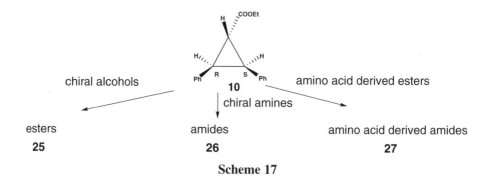

Scheme 17

4.2.1. Chiral-ester-linked auxiliaries

Photoisomerization of the chiral esters **25** were studied in various cation exchanged zeolites [26]. The results are presented in Table 13.

From the above results we can see that up to 55%de can be achieved with menthyl esters **25a–c**, showing that this chiral auxiliary approach leads to moderate diastereoselectivity.

4.2.2. Chiral-amide-linked auxiliaries

With moderate success with esters as chiral auxiliaries, amines were covalently linked with compound **10 (Scheme 11)** to yield chiral amides **26a–c**. The results of the three chiral amides are summarized in Table 14 [15].

Table 13   Asymmetric photoisomerization of chiral ester derivatives, **25**

Substrate	%de[a] in various media					
	Solution	LiY	NaY	KY	RbY	CsY
(–)-menthyl ester, **25a**	4 B	50 B	55 B	30 B	22 B	5 B
(+)-menthyl ester, **25b**	4 B	48 B	50 B	32 B	21 B	4 B
racemic menthyl ester, **25c**	4 B	48 B	58 B	27 B	15 B	1 B
(–)-2-methylbutyl ester, **25d**	0	14 A	19 B	13 B	5 B	2 B
(+)-isomenthyl ester, **25e**	5 B	12 A	32 B	7 A	5 A	2 A
(+)-neomenthyl ester, **25f**	3 B	25 A	40 A	6 A	5 A	3 A
(+)-fenchyl ester, **25g**	4 B	7 B	16 A	4 B	2 B	2 B
(+)-isopinocampheyl ester, **25h**	5 A	2 A	6 B	7 B	13 B	13 B

[a] The first peak to elute from the column is assigned as A and the second as B.

Table 14   Asymmetric photoisomerization of chiral amide derivatives, **26**

Substrate	%de[a] in various media					
	Solution	LiY	NaY	KY	RbY	CsY
3-methyl-2-butylamide, **26a**	2 B	7 A	7 A	3 A	12 B	23 B
1-cyclohexylethylamide, **26b**	2 B	29 B	24 B	26 A	29 A	37 A
1-phenylethylamide, **26c**	2 B	80 B	28 A	14 A	5 A	5 A

[a] The first peak to elute from the column is assigned as A and the second as B.

To our delight, 80% diastereoselectivity was achieved with (+)-1-phenyl ethyl amide **26c** in dry LiY zeolite. Other zeolites gave a moderate diastereoselectivity. Control studies (Figure 6) on compound **26c** showed that the de was 8% when the reaction was carried out with wet LiY zeolite, whereas it is 80% when it was done in dry LiY zeolite. Further, when the reaction was carried out on a silica surface (dry), the %de was 8%, compared to 80% in zeolite. Thus, the zeolite interior plays a crucial role in determining the diastereoselectivity of the reaction. The other two amides, **26a** and **26b**, gave lower diastereoselectivity inside zeolite. In all these cases the diastereoselectivity in solution (hexane–dichloromethane) was very much lower compared to that in the zeolite.

4.2.3.   Amino acid derivatives as chiral auxiliaries

Since amides gave promising results as chiral auxiliaries, amino acid derivatives **27a–f** were also investigated. As high as 83%de was achieved with amino acid derivatives, as shown in Table 15 [15].

It is evident from Table 15 that the cation of the zeolite plays an important role in controlling the diastereoselectivity of the reaction. One interesting fact to note was that

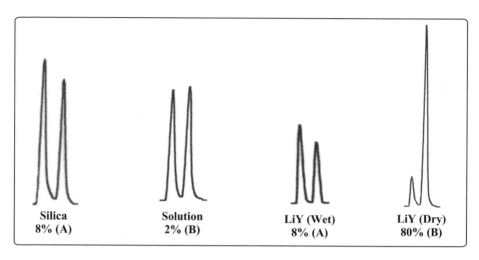

| Silica | Solution | LiY (Wet) | LiY (Dry) |
| 8% (A) | 2% (B) | 8% (A) | 80% (B) |

Figure 6   Control studies with (+)-1-phenylethylamide **26c**.

Table 15   Diasteroselectivity with amino acid derivatives **27a–f** as chiral auxiliaries

Substrate	%de[a] in various media					
	Solution	LiY	NaY	KY	RbY	CsY
L-phenylalanine methyl ester, **27a**	2 A	10 A	32 A	53 A	32 A	6 A
L-phenylalanine ethyl ester, **27b**	1 A	21 A	40 A	44 A	7 A	5 A
L-valine methyl ester, **27c**	2 A	83 B	21 A	80 A	47 A	5 A
L-valine ethyl ester, **27d**	20 A	39 B	31 A	61 A	34 A	0
L-leucine methyl ester, **27e**	3 A	26 B	22 A	46 A	22 A	9 A
L-alanine methyl ester, **27f**	5 B	34 B	22 A	31 A	44 A	5 B

[a] The first peak to elute from the column is assigned as A and the second as B.

for all the compounds except L-phenyl alanine methyl ester, **27a** and ethyl ester, **27b**, the B-isomer predominates in dry LiY whereas the A-isomer predominates in dry NaY, KY, RbY and CsY. When a phenyl substituent was present, as in **27a,b**, the A-isomer predominates from dry LiY to CsY zeolites. This shows that the phenyl ring plays an important role in determining the diastereoselectivity, especially in the case of LiY zeolite. In the absence of a phenyl substituent in the chiral auxiliary part (**27c–f**) the B-isomer always predominates in dry LiY and the A-isomer predominates in dry NaY, KY, RbY and CsY.

Comparing the results of 3-methyl-2-butylamide **26a** and L-valine ethyl ester **27c**, we can conclude that the ester part of the amino acid also plays a crucial role in determining the diastereoselectivity of the reaction. The two substrates differ with respect to the methyl group at the alpha position being replaced by the ester functionality of the amino-acid derivative. This small difference in the functional group increases the %de from 7 to 83 in dry LiY and from 3 to 80 in dry KY. The results presented above suggest that amino-acid derivatives can be effectively used as chiral auxiliaries.

4.3.1. Chiral esters and amides of 1,2-dihydro-2,2-dimethylnaphthalenone 4-carboxylic acid

Compound **19** was also chirally modified to produce chiral esters and amides, **28**, which were then investigated for diastereoselectivity (**Scheme 18**) [18, 20]. Upon solution irradiation, the highest %de values obtained from these substrates were in the range of 4–10%. The results obtained with other chiral esters and amides are summarized in Table 16. These chirally modified substrate molecules produced as high as 81%de with (S)-1-phenylethylamide derivative, **28g**. The effect of cations on the diastereoselectivity for (–) menthyl ester **28b** is shown in Table 17.

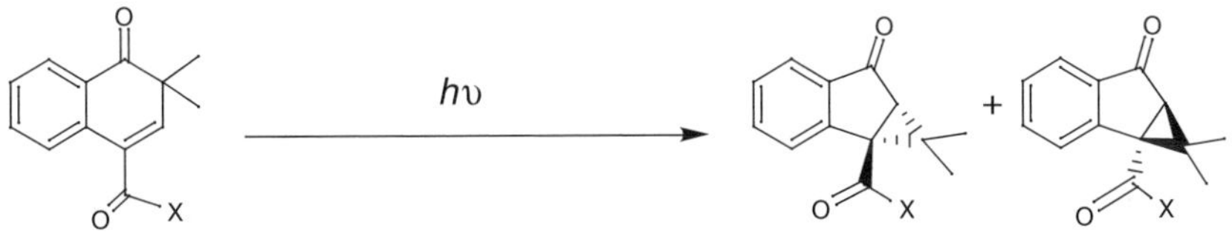

28   **29**   **30**

Scheme 18

4.3.2. Chiral amides of 2,6,6-trimethyl-cyclohexa-2,4-dienone-4-carboxylic acid

Similarly as above, the cyclohexadienone system, (**16** from **Scheme 8**) was chirally modified as amides **31a–c** (**Scheme 19**) [18]. As observed in the case of the naphthalenone systems, the three chirally modified cyclohexadienone systems upon irradiation give moderately high diastereoselectivity within zeolite supercages. It was found

Table 16 Effects of chiral auxiliaries with 1,2-dihydro-2,2-dimethylnaphthalenone 4-carboxylic acid systems, **28**

Substrate	%de[a] in zeolite
(–)-2-methyl-1-butyl ester **28a**	NaY/10 A
(–)-menthyl ester **28b**	KY/60 A
(+)-fenchyl ester **28c**	NaY/57 B
(+)-bornyl amide **28d**	NaY/48 B
(+)-3-methyl-2-butyl amide **28e**	NaY/45 B
(R)-cyclohexylethyl amide **28f**	KY/58 B
(S)-phenylethyl amide **28g**	NaY/81 B

[a] The first peak to elute from the column is assigned as A and the second as B.

Table 17 Effect of cation on diastereoselectivity during photochemistry of **28b**

Medium	%de[a]
hexane solution	9 B
silica gel	2 B
LiY	12 A
NaY	46 A
KY	60 A
RbY	48 A
CsY	10 A

[a] The first peak to elute from the column is assigned as A and the second as B.

Table 18 Auxiliary and cation effect on diastereoselectivity during photochemistry of **31**

Substrate	%de[a] in various zeolites				
	LiY	NaY	KY	RbY	CsY
(+)-phenylethyl amide **31a**	17 A	59 A	53 B	39 B	9 B
(+)-3-methyl-2-butyl amide **31b**	4 A	2 A	9 B	38 B	31 B
(−)-1-methoxy-2-propyl amide **31c**	40 A	73 A	1 A	25 B	20 B

[a] The first peak to elute from the column is assigned as A and the second as B.

that diastereoselectivity was dependent on the chiral auxiliaries used and the cations present in the zeolite (Table 18).

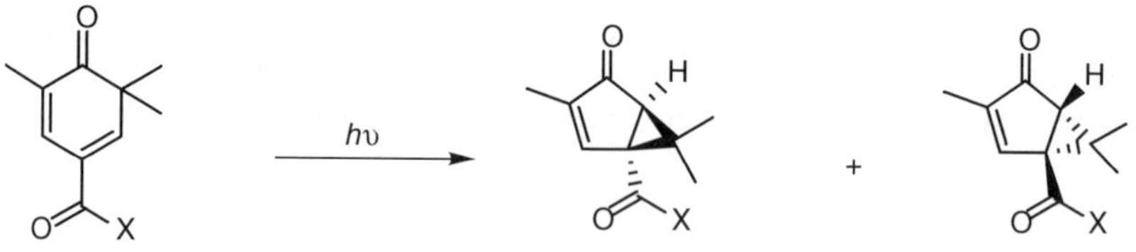

Scheme 19

4.4. Cyclization of tropolone ethers

Irradiation of the (S)-tropolone 2-methyl butyl ether in solution yields a 1:1 mixture of diastereomeric products (Figure 7). Clearly, in solution the presence of the chiral auxiliary in proximity to the reactive center has no influence on the product stereochemistry. On the other hand, the same molecule when irradiated within NaY gave the product in ~53% diastereomeric excess (Figure 7) [23, 27]. The smaller and controlled space provided by the zeolite supercage apparently has forced the chiral center to establish communication with the reaction site. Considering that the chiral center has only alkyl

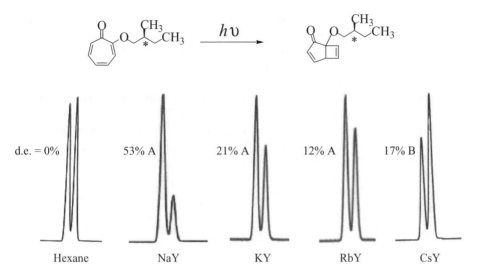

Figure 7 Results of studies on the photocyclization of (S)-tropolone 2-methyl butyl ether within various cation exchanged Y zeolites. The HPLC traces clearly illustrate the dependence of d.e. on the cation. Note the isomer B increases with the increasing size (and decreasing charge density) of the cation. The % d.e. (diastereomeric excess) and the isomer enhanced are shown on the HPLC traces.

groups one would not expect any special interactions to develop between the chiral auxiliary and the reaction centers within a zeolite. It is the small 'reaction cavity' that has enforced a certain amount of discipline on the reacting molecule. As shown in Figure 7, the cation plays a very important role. As the cation size increases (or the binding ability decreases) the extent of diastereoselectivity decreases. By the time one reaches Cs$^+$ the isomer that is enhanced is switched from A to B. This is the first clear indication that the cation controls the extent and the nature of selectivity.

5. Bimolecular Approach

By restricting the photoreaction to those cages in which the reactant and the chiral inductor are together, we can overcome the problem of random distribution of molecules

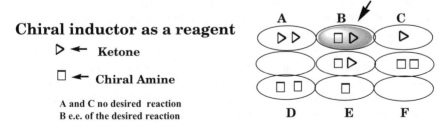

Figure 8 Possible distributions of ketone and chiral amine in bimolecular approach.

181

leading to racemic products. Towards this, we have devised a strategy that would limit the photoreaction of interest only to the reactant molecules that are next to the chiral inductor (Figure 8). This eliminates the possibility of formation of the product of interest in cages that do not contain both the chiral inductor and the reactant molecule.

The photoreaction we have investigated in this context is the electron-transfer-initiated intermolecular hydrogen abstraction reaction of carbonyl compounds [28]. It is well documented that amines transfer electrons to the excited carbonyls to generate carbonyl radical anions that react further to give benzylic alcohols (**Scheme 20**). Under the conditions we have employed, one of the guest molecules (e.g., ephedrine) assumes the dual role of chiral inductor and electron donor.

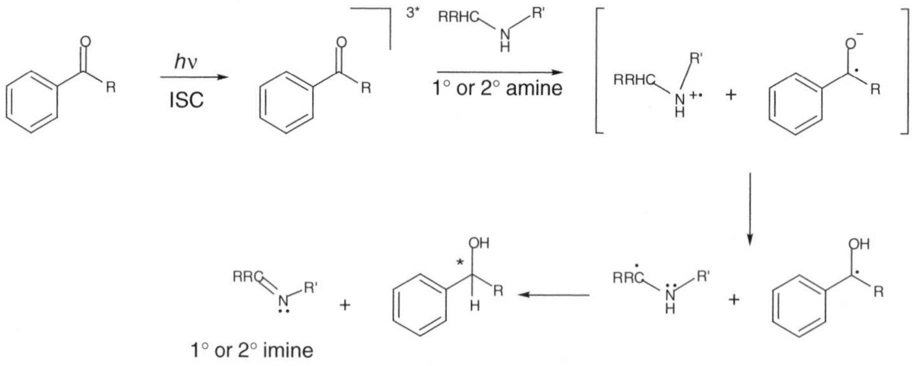

Scheme 20

5.1. *Photoreduction of aryl alkyl ketones*

The photochemistry of phenyl cyclohexyl ketone **34** has been used to establish the above concept [29]. As expected on the basis of solution behavior, irradiation of a hexane slurry of **34** included within zeolite NaY gave only intramolecular hydrogen abstraction product **35**. However, irradiation of **34** included within ephedrine-, pseudoephedrine-, or norephedrine-modified NaY gave intermolecular hydrogen abstraction product **36**, in addition to **34** (**Scheme 21**) [30].

Product **36** is the result of electron transfer from the amino group of the chiral inductors, as it was not formed in their absence. The absence of **36** when ephedrine hydrochloride was used as the chiral inductor supports this view. Further, **36** was not formed when the chiral inductor was (–)-diethyl tartrate, which does not contain an amino group. If the role of the amine is to serve as an electron donor, the ratio of inter- versus intramolecular hydrogen abstraction products (**36** to **35**) should depend on the electron-donating ability of the chiral inductor. Since secondary amines are better electron donors than primary, it is no surprise that the **36/35** ratio is higher for pseudoephedrine (5) than for norephedrine (0.2). Also, since intermolecular reduction takes place only when the electron-donating amine is near the electron-accepting ketone, the **36/35** ratio should depend on the loading level of the amine. This is also observed, as the ratio increases from 1.3 to 5 when the loading level (molecules per supercage) is increased from 0.2 to 1.

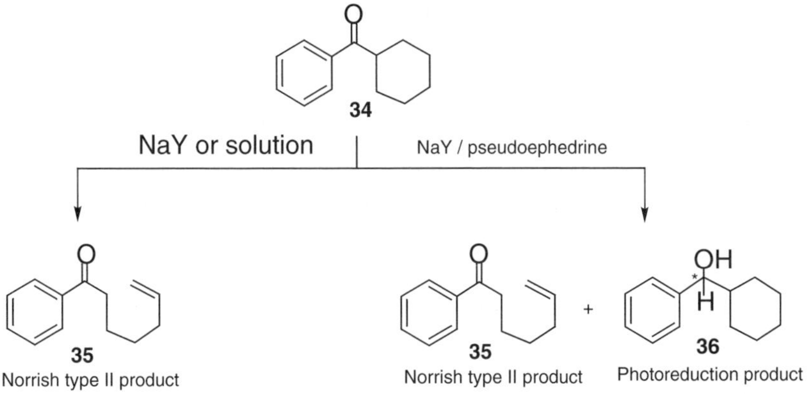

Scheme 21

Table 19 Enantiomeric excesses obtained with various chiral inductors included in NaY

Chiral inductor	% ee[a] in **36**	Chiral inductor	%ee[a] in **36**
(–)-norephedrine	68 B	(–)-ephedrine	13 B
(+)-norephedrine	67 A	(+)-ephedrine	16 A
(–)-pseudoephedrine	35 B	D-valinol	10 A
(+)-pseudoephedrine	33 A	L -valinol	9 B
(+)-2-amino-3-methoxy- 1-phenyl-1-propanol	30 B	L-methyl benzylamine	8 B
(+)-2-amino- 1-phenyl-1,3-propanediol	54 B	L-phenylalaninol	13 A
(–)-diaminocyclohexane	28 - A	L-phenylglycinol	11 - B

[a] The first peak to elute from the column is assigned as A and the second as B.

Although the **36/35** ratio depends on the loading level of amine, the %ee of **36** is observed to be independent of the loading level of the amine. The enantioselectivity with various chiral inductors are summarized in Table 19. Of the various chiral inductors tested, the best results were obtained with norephedrine (68%ee). The use of (+)-norephedrine afforded the optical antipode of the photoproduct produced by the use of (–)-norephedrine, indicating that the system is well behaved. Similar to our observations with other systems discussed in this chapter, the %ee obtained was dependent on the water content of the zeolite. When the NaY/norephedrine complex was intentionally made 'wet' by adsorption of water, the %ee dropped from 68 to 2.

Cations affect the enantioselectivity (Table 20), probably by influencing the arrangement of the molecules in the supercage. Another interesting aspect is the effect of temperature on the %ee observed during the photoreduction. In general, electron transfer reactions undergo little or no change with change in temperature. The same is observed with the photoreduction of ketones by amines. The %ee remains almost constant with temperature variation (Table 21).

Table 20   Effect of cation on the enantioselectivity during the reduction of **34** by
(+) norephedrine

Zeolite	%ee[a]
LiY	45 A
NaY	68 A
KY	30 A
RbY	10 A

[a] The first peak to elute from the column is
assigned as A and the second as B.

Table 21   Effect of temperature on the
enantioselectivity during the reduction of **34**
by (+) norephedrine in NaY

Temperature /°C	% ee[a]
RT	68 A
20	67 A
0	63 A
−20	65 A
−40	63 A

[a] The first peak to elute from the column is
assigned as A and the second as B.

The strategy presented above with phenyl cyclohexyl ketone was established to be general by investigating a number of aryl alkyl ketones [31]. The best cases of %ee are summarized in Figure 9.

While the %ee obtained in this study is the highest thus far reported for the photo-reduction of any achiral ketone, we do not fully understand the mechanism of chiral induction. Despite the entire reaction occurring within chirally modified cages, the %ee

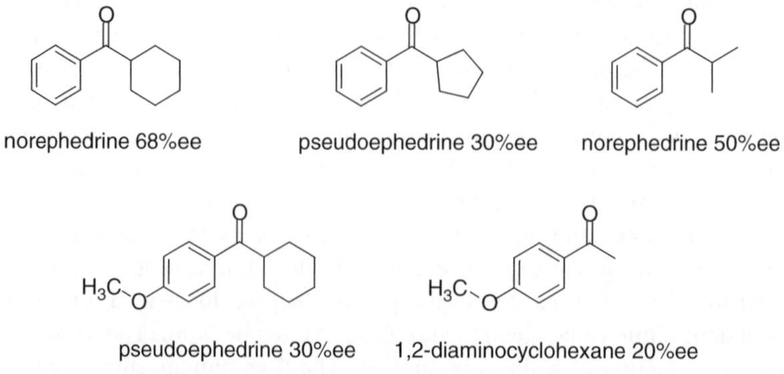

norephedrine 68%ee   pseudoephedrine 30%ee   norephedrine 50%ee

pseudoephedrine 30%ee   1,2-diaminocyclohexane 20%ee

Figure 9   Best cases of chiral induction with various aryl alkyl ketones.

is not quantitative. This is most likely related to the multi-step nature of the reaction, which involves at least two distinct intermediates.

5.2. *Photoreduction of asymmetric benzophenones*

We have also adopted the above strategy and achieved moderate chiral induction during the photoreduction of 2-ethoxybenzophenone **37** [30]. This molecule gives intramolecular cyclization product **38** as the only product in solution as well as within NaY. However, in NaY in the presence of chiral amines, intermolecular reduction product **39**, in addition to **38**, was obtained (**Scheme 22**). More importantly, with pseudoephedrine and (1*R*,2*R*)-diaminocyclohexane as chiral inductors, moderate %ee was obtained on the product **39** (39 and 51%, respectively). Several other substituted benzophenones were also investigated and the best cases of %ee are given in Figure 10 [32].

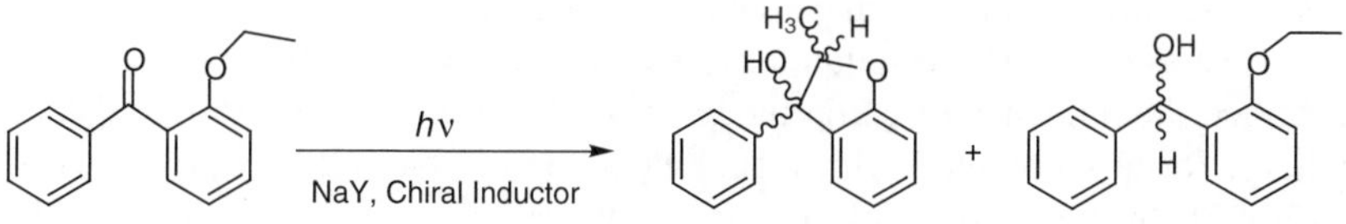

Scheme 22

pseudoephedrine 43% ee

1,2 diaminocyclohexane 44% ee

ephedrine 45% ee

ephedrine 23% ee

pseudoephedrine 22% ee

ephedrine 34% ee

pseudoephedrine 35% ee

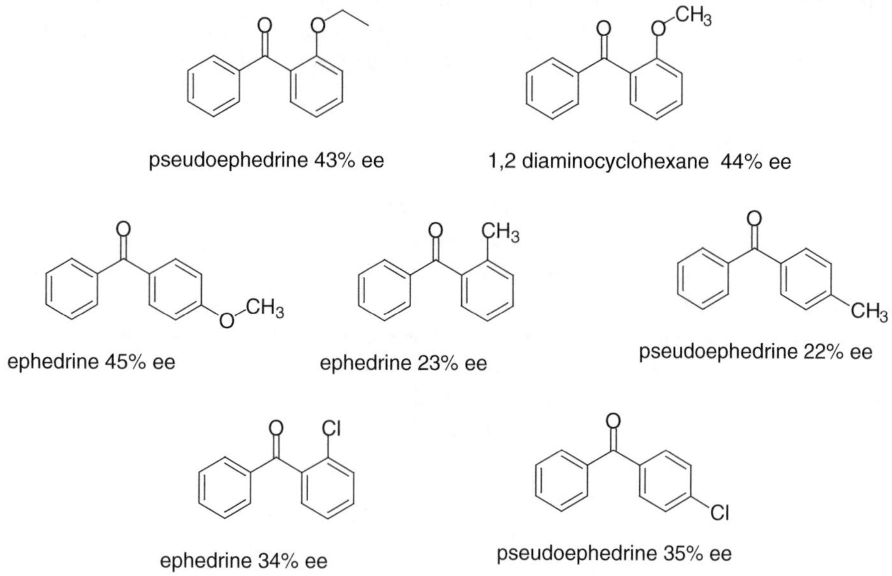

Figure 10 Best cases of chiral induction in the intermolecular photoreduction of various benzophenones.

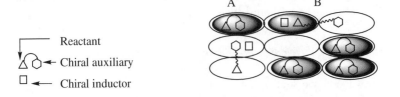

Reactant

Chiral auxiliary

Chiral inductor

Figure 11   Chiral auxiliary and chiral indutor combined to agument chiral induction during a photoreaction.

6.   Chiral Inductor and Chiral Auxiliary Strategies Combined

As shown in Figure 11 even when the chiral inductor is covalently attached to the reactant there is a possibility that the two may remain in different cages (type B in Figure 11). These types of cages could be the source of less than 100% d.e. To improve this situation we examined the influence of chiral inductors in the systems that already contains chiral auxiliary. Results obtained with tropolone 2-methyl butyl ether are shown in Figure 12 [20,27]. In the absence of ephedrine, diastereomer A is obtained in 53% diastereomeric excess. When (–)-ephedrine was used as the chiral inductor the same isomer was enhanced to the extent of 90%. On the other hand, (+)-ephedrine favors the B diastereomer to the tune of 70% diastereomeric excess. These observations are truly remarkable. High d.e. has been achieved through the use of two chiral inducers and a confined space. The importance of this result becomes more apparent when one recognizes that irradiation in solution of the same compound in presence of ephedrine gave a 1:1 diastereomeric mixture. Zeolite is essential to achieve the high d.e.

7.   Summary and Future

Four methods have been explored to examine the use of zeolites as reaction media for chiral induction during photochemical reactions. Consistently higher e.e. or d.e. has been obtained within zeolites than in solution. The confined space and the cations present within zeolites are believed to be responsible for the chiral induction within zeolites. The size of the reaction vessel in which the reactant molecule is housed during irradiation matters. Clearly molecular-sized reaction vessels are better than the conventional macro-sized containers. The e.e. obtained within zeolites is still not quantitative, suggesting that there are factors yet to be optimized. Experiments are underway along this direction.

Acknowledgement

We thank the National Science Foundation, USA for support of the work presented in this chapter.

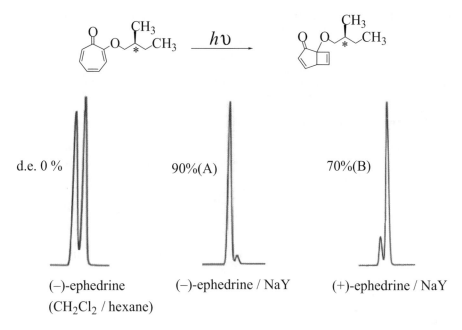

Figure 12 Results of studies on the photocyclization of (S)-tropolone 2-methyl butyl ether within chiraly modified NaY zeolite. The % d.e. (diastereomeric excess) and the isomer enhanced are shown on the HPLC traces. The non-identical reversal of the product enantiomer by the antipodes of the chiral inductor ephedrine is due to the fact that not all molecules are in proximity to a chiral inductor. Chiral induction also occurs due to the chiral auxiliary present in the molecule.

References

1. (a) S. C. STINSON, *Chem. Eng. News.*, Sep 21, (1998) 83. (b) S. C. STINSON, *Chem. Eng. News.*, Oct 20, (1997) 38 (c) M. AVALOS, R. BABIANO, P. CINTAS, J. L. JIMENEZ and J. C. PALACIOS, *Chem. Rev.* **98** (1998) 2391. (d) A. RICHARDS and R. MCCAGUE, *Chemistry & Industry* (1997) 422. (e) R. NOYORI, "Asymmetric Catalysis in Organic Synthesis", (Wiley-Interscience, New York, 1994) (f) O. CERVINKA, "Enantioselective Reactions in Organic Chemistry", (Ellis Horwood, London, 1995).

2. (a) Y. INOUE, *Chem. Rev.* **93** (1992) 741. (b) H. RAU, *Chem. Rev.* **83** (1983) 535. (c) J. P. PETE, *Adv. Photochem.* **21** (1996) 135. (d) S. R. L. EVERITT and Y. INOUE, in "Molecular and Supramolecular Photochemistry", Vol. 3, edited by V. Ramamurthy and K. S. Schanze, (Marcell Dekker, New York, 1999). p. 71. (c) H. BUSCHMANN, H. D. SCHARF, N. HOFFMANN and, P. ESSER, *Angew. Chem. Int. Ed. Engl.* **30** (1991) 477.

3. V. RAMAMURTHY and K. VENKATESAN, *Chem. Rev.* **87** (1987) 433.

4. M. VAIDA, R. POPOVITZ-BIRO, L. LESEROWITZ and M. LAHAV in "Photochemistry in Organized and Constrained Media", edited by V. Ramamurthy, (VCH, New York, 1991) pp. 247–302.

5. (a) J. N. GAMLIN, R. JONES, M. LEIBOVITCH, B. PATRICK, J. R. SCHEFFER and J. TROTTER, *Acc. Chem. Res.* **29** (1996) 203. (b) M. LEIBOVITCH, G. OLOVSSON, J. R. SCHEFFER and J. TROTTER, *Pure & Appl. Chem.* **69** (1997) 815.

6. (a) F. TODA, *Acc. Chem. Res.* **28** (1995) 480.

    (b) K. TANAKA and F. TODA, *Chem. Rev.* **100** (2000) 1025.

7. G. DESIRAJU, "Crystal Engineering, The Design of Organic Solids", (Elsevier, Amsterdam, 1989).

8. (a) D. W. BRECK, "Zeolite Molecular Sieves: Structure, Chemistry and Use", (John Wiley and Sons, New York, 1974). (b) A. DYER, "An Introduction to Zeolite Molecular Sieves", (John Wiley and Sons,

New York, 1988). (c) H. VAN BEKKUM, E. M. FLANIGEN and J. C. JANSEN, "Introduction to Zeolite Science and Practice", (Elsevier, Amsterdam, 1991).

9. (a) S. M. STALDER and A. P. WILKINSON, *Chem. Mater.* **9** (1997) 2168. (b) R. SZOSTAK, "Handbook of Molecular Sieves", (Van Nostrand Reinhold, New York, 1992). (c) R. W. TSCHERNICH, "Zeolites of the World", (Geoscience Press Inc., Phoenix, 1992). (d) J. M. NEWSAM, M. M. J. TREACY, W. T. KOETSIER and C. B. DE GRUYTER, *Proc. Royal Soc. Lond. A* **420** (1988) 375. (e) M. W. ANDERSON, O. TERASAKI, T. OHSUNA, A. PHILIPPOU, S. P. MACKAY, A. FERREIRA, J. ROCHA and S. LIDIN, *Nature* **367** (1994) 347. (f) D. E. AKPORIAYE, *J. Chem. Soc., Chem. Commun.* (1994) 1711.

10. (a) P. J. WAGNER and B. S. PARK, *Org. Photochem.* **11** (1991) 227. (b) N. C. YANG and D. -H. YANG, *J. Am. Chem. Soc.* **80** (1958) 2913. (c) P. J. WAGNER, *Acc. Chem. Res.* **22** (1989) 83. (d) J. C. SCAIANO, *Acc. Chem. Res.* **15** (1982) 252.

11. M. LEIBOVITCH, G. PLOVSSON, G. SUNDARABABU, V. RAMAMURTHY, J. R. SCHEFFER and J. TROTTER, *J. Am. Chem. Soc.* **118** (1996) 1219.

12. G. SUNDARABABU, M. LEIBOVITCH, D. R. CORBIN, J. R. SCHEFFER and V. RAMAMURTHY, *J. Chem. Soc., Chem. Comm.* (1996) 2159.

13. N. ARUNKUMAR, A. JOY and V. RAMAMURTHY, unpublished results.

14. (a) G. S. HAMMOND and R. S. COLE, *J. Am. Chem. Soc.* **87** (1965) 3256. (b) A. FALJONI, K. ZINNER and R. G. WEISS, *Tetrahedron Lett.* **13** (1974) 1127. (c) C. OUANNES, R. BEUGELMANS and G. ROUSSI, G. J. Am. *Chem. Soc.* **95** (1973) 8572. (d) L. HORNER and J. KLAUS, *Liebigs. Ann. Chem.* (1979) 1232. (e) Y. INOUE, N. YAMASAKI, H. SHIMOYAMA and A. TAI, *J. Org. Chem.* **58** (1993) 1785.

15. J. SIVAGURU and V. RAMAMURTHY, unpublished results.

16. (a) H. E. ZIMMERMAN and T. W. FLECHTNER, *J. Am. Chem. Soc.* **92** (1970) 6931. (b) H. E. ZIMMERMAN and C. M. MOORE, *J. Am. Chem. Soc.* **92** (1970) 2023.

17. J. GRIFFITHS and H. HART, *J. Am. Chem. Soc.* **90** (1968) 5296

18. S. UPPILI and V. RAMAMURTHY, unpublished results.

19. H. HART and R. K. MURRAY, *J. Org. Chem.* **35** (1970) 1535.

20. A. JOY, S. UPPILI, M. R. NETHERTON, J. R. SCHEFFER and V. RAMAMURTHY, *J. Am. Chem. Soc.* **122** (2000) 728.

21. (a) W. G. DAUBEN, K. KOCH, S. L. SMITH and O. L. CHAPMAN, *J. Am. Chem. Soc.* **85** (1963) 2616. (b) O. L. CHAPMAN, *Adv. Photochem.* **1** (1963) 302.

22. H. TAKASHITA, M. KUMAMOTO and I. KOINO, I. Bull. *Chem. Soc. Jap.* **53** (1980) 1006. (b) F. TODA and K. TANAKA, *J. Chem. Soc. Chem. Commun.* (1986) 1429. (c) M. KAFTORY, M. YAGI, K. TANAKA, and F. TODA, *J. Org. Chem.* **53** (1988) 4391. (d) M. KAFTORY, F. TODA, K. TANAKA and M. YAGI, *Tetrahedron* **186** (1990) 167.

23. (a) A. JOY, J. R. SCHEFFER, D. R. CORBIN and V. RAMAMURTHY, *J. Chem. Soc. Chem. Commun.* (1998) 1379. (b) A. JOY, D. R. CORBIN and V. RAMAMURTHY in "Proceedings of twelvth International Zeolite Conference", edited by M. M. J. Traecy, B. K. Marcus, B. E. Bisher and J. B. Higgins, (Materials Research Society, Warrendale, PA, (1999)) p. 2095. (c) ABRAHAM JOY, JOHN R. SCHEFFER and V. RAMAMURTHY, *Org. Lett.* **2** (2000) 119.

24. A. JOY, R. J. ROBBINS, K. PITCHUMANI and V. RAMAMURTHY, *Tetrahedron Lett.* (1997) 8825.

25. (a) S. JAYARAMAN, S. UPPILI, A. NATARAJAN, A. JOY, K. C. W. CHONG, M. R. NETHERTON, A. ZENOVA, J. R. SCHEFFER and V. RAMAMURTHY, *Tetrahedron Lett.* **41** (2000) 8231. (b) N. ARUNKUMAR, J. R. SCHEFFER and V. RAMAMURTHY, unpublished results.

26. E. CHEUNG, K. C.W. CHONG, S. JAYARAMAN, V. RAMAMURTHY, J. R. SCHEFFER and JAMES TROTTER, *Organic Lett.* **2** (2000) 2801.

27. A. JOY and V. RAMAMURTHY, *Chemi. Euro. Journal* **6** (2000) 1287.

28. S. G. COHEN, A. PAROLA and G. H. PARSONS, *Chem. Rev.* **73** (1973) 141.

29. (a) F. D. LEWIS, R. W. JOHNSON, and D. E. J. JOHNSON, *J Am. Chem. Soc.* **96** (1974) 6090. (b) J. H. STOCKER, and D. H. KERN, *J. C. S. Chem. Commun.* (1969) 204.

30. J. SHAILAJA, K. J. PONCHOT and V. RAMAMURTHY, *Org. Lett.* **2** (2000) 937.

31. J. SHAILAJA and V. RAMAMURTHY, unpublished results.

32. K. PONTCHOT and V. RAMAMURTHY, unpublished results.

188

Solid-State Bimolecular Photoreactions in Two-Component Molecular Crystals

Hideko Koshima

Department of Applied Chemistry, Faculty of Engineering, Ehime University,
Matsuyama 790-8577, Japan
E-mail:koshima@en3.ehime-u.ac.jp

1. Introduction

Solid-state organic photochemistry is a promising field since solution photochemistry has been comprehensively developed over the last few decades. Crystals in which molecules are regularly arranged in three-dimensions have the possibility of more highly selective reactivity than in solution, as well as specific reactions that are different from those observed in solution. Hence, such crystals are attractive as a reaction medium.

Since the study of [2 + 2] photocycloaddition of *trans*-cinnamic acids by Cohen and Schmidt in 1964 [1], a large number of solid-state photoreactions have been reported [2–11]. However, most of these reactions were unimolecular (intramolecular) photoreactions occurring in one-component crystals or inclusion complexes. Bimolecular (intermolecular) photoreactions were very few in number, and almost limited to [2 + 2] photocycloadditions in one-component crystals, mixed crystals or two-component crystals from similar molecules. One of the reasons why solid-state photochemistry has not developed into a versatile area of organic photochemistry is due to the scarcity of solid-state bimolecular photoreactions, in comparison to solution photochemistry. In order to expand the scope of solid-state organic photochemistry, it is therefore obvious that solid-state bimolecular photoreactions between two different organic compounds should be developed.

We have studied the photochemistry of two-component molecular crystals by focusing on the preparation of two-component crystals over the past decade and this has led to us obtaining fruitful results [12,13]. These include highly selective photodecarboxylation using stoichiometric sensitization and a coupling reaction leading to an absolute asymmetric synthesis. An important advantage in utilizing a two-component crystal as a reactant is that even if a component molecule has no photoreactivity by itself, new reactivity can be induced by combining an electron donor and an electron acceptor in the crystal and exploiting photoinduced electron transfer. The preparation of photoreactive two-component crystals is a key point of this study.

In this chapter, recent selected studies of our own and other researchers are reviewed, with a particular emphasis on the correlation between reactivities and X-ray crystal structures. The contents are constructed from preparation of two-component crystals, bimolecular photoreactions in simple polycrystalline mixtures, and various types of bimolecular photoreactions in two-component crystals such as hydrogen abstraction, addition, [2 + 2] cycloaddition, photodecarboxylation and absolute asymmetric synthesis. I hope this chapter will explain clearly the features of bimolecular photoreactions in

F. Toda (ed.), Organic Solid-State Reactions, 189–268.

$$\text{A} \; + \; \text{B} \quad \xrightarrow{\quad \textit{crystallization} \quad} \quad \text{A·B} \quad \xrightarrow{\quad h\nu \quad} \quad \text{P}$$

the crystalline state and also create a wider interest in the photochemistry of two-component crystals.

2. Preparation of Two-component Molecular Crystals

In order to carry out bimolecular photoreactions between two different organic molecules in the solid state, it is first necessary to prepare two-component molecular crystals containing two different reactants. Two requirements in this preparation must be simultaneously satisfied, namely formation of the two-component crystal and also induction of photoreactivity. The preparation of two-component crystals is a type of crystal engineering [14,15], and is quite different from the more familiar photochemistry in the solution phase.

Two-component crystals can be prepared by the following two methods:

(1)   crystallization from solution of a mixture of two compounds by means of evaporation or cooling.
(2)   melting-resolidification, which involves the heating of a mixture of two crystalline compounds to a homogeneous melt, followed by resolidification on cooling to give a polycrystalline solid.

The crystals obtained are classified into four types: (a) simple polycrystalline mixture of both components, (b) stoichiometric two-component molecular crystal (co-crystal), (c) mixed crystal (substitutional solid solution), and (d) amorphous solid (Figure 2.1). The target of this chapter is to obtain type (b), i.e. a genuine pure two-component crystal whose crystal structure is different from those of both components. When combining similar molecules, the type (c) mixed crystal (which is isomorphous with the two components) is sometimes formed, but such mixed crystals will not be dealt with here.

The four types of crystals can be distinguished by characterization in the solid state using several physical methods, such as differential scanning calorimetry (DSC), infrared (IR) spectrometry, powder X-ray diffractometry (PXD), and solid state ^{13}C nuclear magnetic resonance (NMR). Sometimes construction of the phase diagram is useful; the example of the two-component crystal of duroquinone and durene is shown in Figure 4.1; that of a simple mixture of benzophenone and benzhydrol in Figure 3.1. When a new two-component crystal is first obtained, it is most important to prepare a high-quality single crystal and perform the X-ray crystallographic analysis in order to elucidate the correlation between its reactivity and the molecular arrangement in the crystalline state.

Factors that may facilitate the formation of two-component crystals are hydrogen bonding, salt formation, charge transfer (CT) interaction, and hydrophobic interaction. Hydrogen bonding is the most efficient; this is understandable from the relatively large hydrogen bond energy for interactions such as X–H•••Y (X, Y = N, O, P, S, halogen),

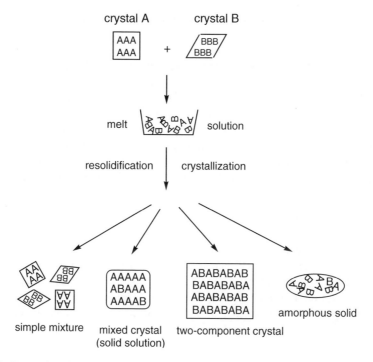

Figure 2.1 Preparation of two-component crystals from two different organic compounds.

which lies in the range 8–30 kJ mol^{-1} [16]. Much weaker hydrogen bonding such as the C–H•••π interaction is also utilized. A salt is formed in the case of acid-base interactions, such as components with carboxylic acid and amine functionalities.

Another important consideration for the preparation of two-component crystals is the induction of photoreactivity. As a result of our experiences, we found that even if one particular component molecule is inert, photoreactivity may be induced via photoinduced electron transfer by combining an electron acceptor and an electron donor in the crystal. Hence combinations such as acridine and phenothiazine (Section 5.1) or aza aromatic compounds and aralkylcarboxylic acids (Section 8) within two-component crystals result in effective photoreactions. CT crystals have been studied so much in the last few decades that a large number are already known and these may be considered as typical organic crystals. Although it had been thought that strong CT crystals have no photoreactivity due to fast back-electron transfer, several CT crystals of moderate strength do have photoreactivities, as shown in Sections 5.2, 6.3, 6.4, 8.2, 8.3 and 9.2. Re-examination of known CT crystals may be useful in uncovering new solid-state bimolecular photoreactions. In addition, the combination of a sensitizer and a non-photoreactive molecule can also help induce photoreactivity.

Thus, a possible advantage in utilizing two-component crystals as the reaction medium may be the induction of photoreactivity. Furthermore, numerous combinations of two different compounds can be employed in developing various types of reactions.

3. Photochemical Reactions in Simple Polycrystalline Mixtures

First, solid-state bimolecular photoreactions are introduced that occur in simple polycrystalline mixtures of two different organic compounds [13]. Several types of photoreactions are included such as hydrogen abstraction [17,18], photoaddition of NH compounds to aromatic and olefinic double bonds [19–22], condensation reaction via photoinduced electron transfer [23] and [2 + 2] photocycloaddition [24,25]. These reactions occur at the interfaces between microcrystals of both components, while reactions in two-component crystals take place within in the crystal lattice. The product selectivities for the photoreactions in solid mixtures are sometimes different from those in solution photoreactions, but are necessarily lower than those in two-component crystals.

3.1. Hydrogen abstraction

It is well known that the most typical photochemical reaction is a hydrogen abstraction by excited benzophenone **1** to give benzopinacol **3** as a product. In the solution photoreaction, 2-propanol is commonly used as the hydrogen donor. Similarly, solid-state hydrogen abstraction occurs by utilizing various solid hydrogen donors such as benzhydrol **2** or durene **4** [17]. **Scheme 3.1** shows typical selected reactions. All the combinations prepared by melting and resolidification were simple polycrystalline mixtures of

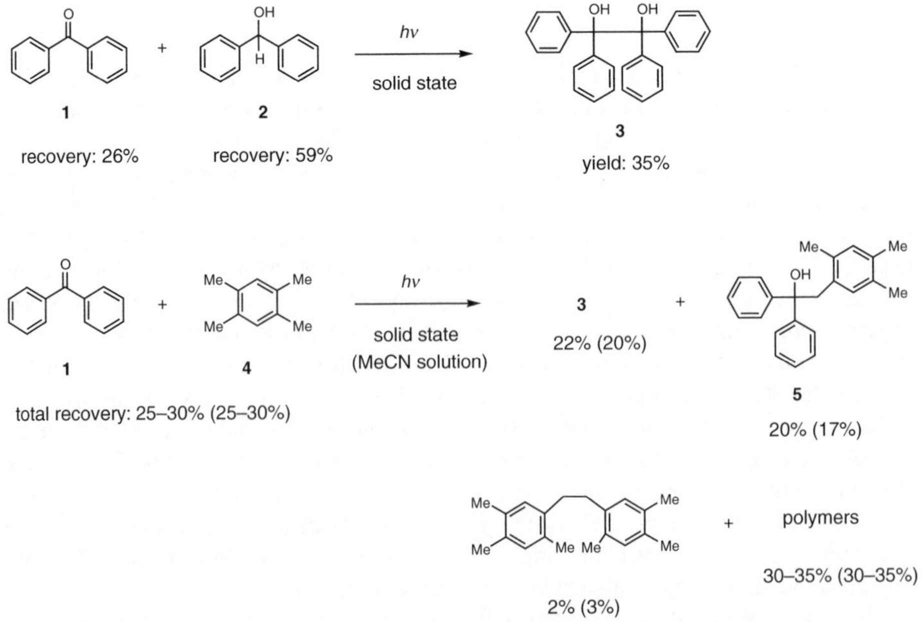

Scheme 3.1   Hydrogen abstraction reaction between benzophenone and donors in the polycrystalline mixtures.

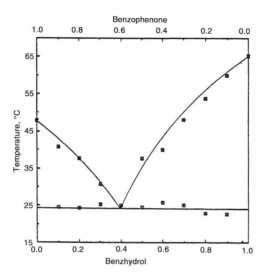

Figure 3.1 Phase diagram of the simple crystalline mixtures of benzophenone and benzhydrol. *Source*: from ref. 17 with permission. © 1995 Elsevier Science.

both components. The phase diagram of benzophenone **1** and benzhydrol **2** constructed by DSC and PXD revealed the typical simple mixture having a eutectic point at 24 °C (Figure 3.1). Irradiation of the 1:1 simple solid mixture of **1** and **2** with a high-pressure mercury lamp through Pyrex glass gave benzopinacol **3** as the sole product in fairly good yield (**Scheme 3.1**). The product distribution in the photoreaction between **1** and durene **4** in the solid state is almost the same as that in acetonitrile solution. This affords benzopinacol **3** and the cross-coupling product **5** as the two main products, indicating that the molecular arrangement of the substrates in the polycrystalline state of **1** and **4** may be random and similar to that in solution.

Another example is hydrogen abstraction by quinones from polymethylbenzenes [18]. The crystals incorporate quinones such as 1,4-benzoquinone **6**, 2,5-xyloquinone **7** and duroquinone **8**; and polymethylbenzenes such as durene **9**, pentamethylbenzene **10** and hexamethylbenzene **11** (**Scheme 3.2**). Nine solid mixtures of **6–8** with **9–11** were prepared by the melting-resolidifying process, and characterized by PXD, DSC, IR spectroscopy and the construction of phase diagrams. Only the crystals from **8** and **9** were found to be 2:1 two-component molecular crystals (see Section 4.1), and the other eight crystals were simple polycrystalline mixtures of both components.

Irradiation of the solid mixtures **6/9**, **6/10** and **8/10** caused intermolecular hydrogen abstraction resulting in the corresponding hydroquinones in 29%, 36% and 23% yield, respectively. In the case of the mixed crystals containing xyloquinone **7**, photodimers of **7** were the major product and the corresponding hydroquinone was formed in only minute amounts. Product selectivities in the solid state were different from those in acetonitrile solution. For example, the solid mixture **7/9** gave the corresponding hydroquinone in 43% yield in acetonitrile solution, but yielded photodimers of **7** in the solid state.

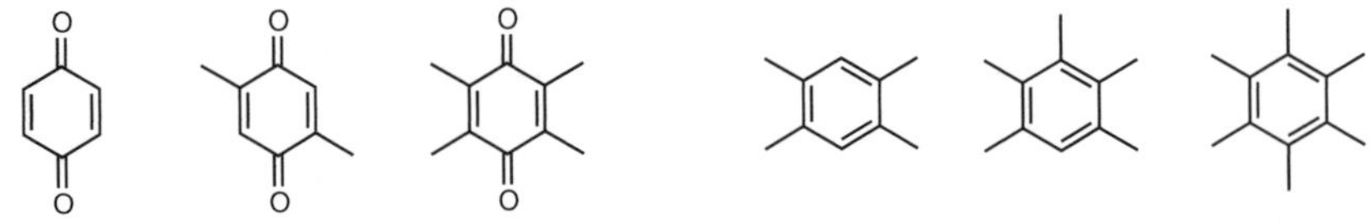

6   7   8   9   10   11

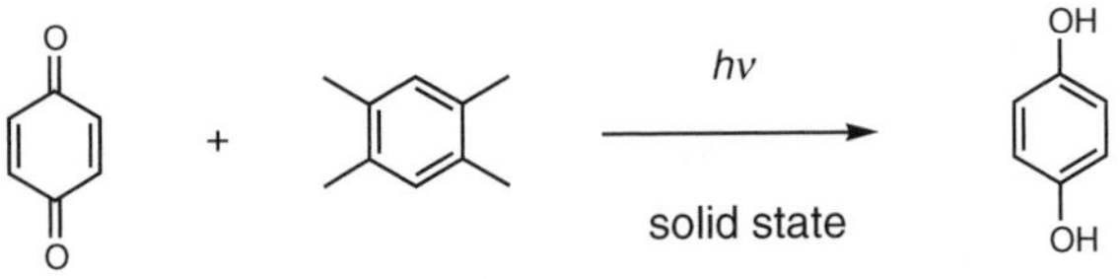

Scheme 3.2   Hydrogen abstraction between benzoquinones and polymethylbenzenes in the polycrystalline mixtures.

3.2.   Photoaddition of NH compounds to aromatics and olefinic double bonds

Solid-state photoadditions occur between NH compounds such as indole, carbazole and diphenylamine to aromatic and olefinic compounds such as phenanthrene and *trans*-stilbene in the simple polycrystalline mixtures to give the corresponding adducts with an N–C bond as the main products [19–22]. Typical examples are summarized in **Scheme 3.3**. Application of the melting-resolidifying process to all the 1:1 combinations of both components afforded simple mixtures of their microcrystals. Irradiation of these crystals gave the corresponding addition products **14**, **19** and **20–24** but the yields were not always satisfactory. However, as shown for the formation of **19** from carbazole **15** and *trans*-stilbene **18**, or from phenothiazine **16** and *trans*-stilbene **18**, reasonably good yields were obtained when an excess of *trans*-stilbene was used for the preparation of the polycrystalline mixtures [21]. The yields of the products also depend on the ratio of the two substrates in the simple mixtures of diphenylamine **17** with phenanthrene **13** or *trans*-stilbene **18** [22].

In contrast, solution-phase photoreactions of these substrates are very different from the solid-state photoreactions. For instance, photolysis of indole **12** and *trans*-stilbene **18** in acetonitrile at low concentrations (0.05 M:0.05 M) gave no photoadducts, but *cis*-stilbene and phenanthrene were major products as the result of isomerization of *trans*-stilbene followed by cyclization.

These photoaddition reactions have been interpreted to occur via initial electron transfer, followed by coupling of the intermediate free radicals, as illustrated in **Scheme 3.4** for NH heteroaromatics (Ar–NH) and *trans*-stilbene **18** [21].

3.3.   Condensation

Crystalline mixtures of arylacetic acids **25a–d** and 1,2,4,5-tetracyanobenzene (TCNB) **26** prepared by evaporating their 1:1 solutions in acetonitrile were shown to be simple polycrystalline mixtures of each component due to no appearance of any new peaks in

194

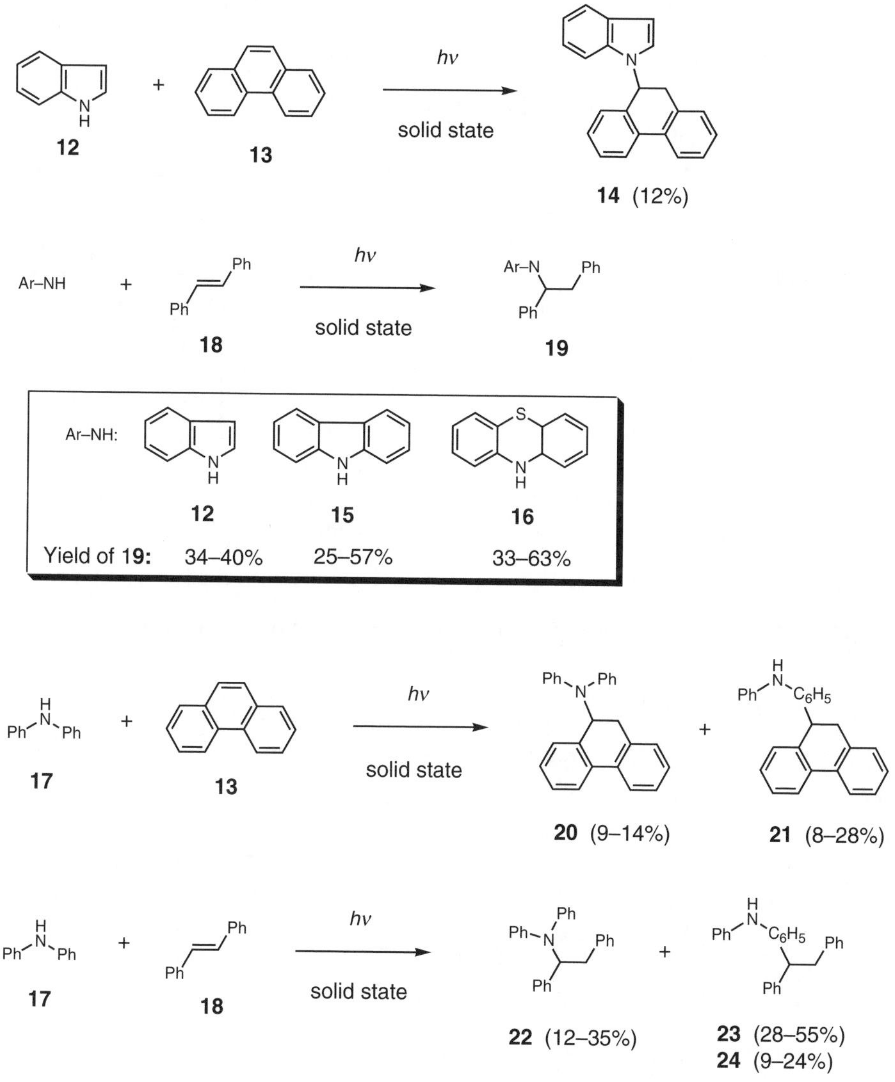

Scheme 3.3 Photoaddition of NH compounds to phenanthrene and *trans*-stilbene in the polycrystalline mixtures.

their PXD patterns and IR spectra. Irradiation of the pulverized mixtures of arylacetic acids **25a–d** and TCNB **26** resulted in decarboxylation followed by dehydrocyanative condensation to give the corresponding arylmethanes **27**, monoarylmethyltricyanoben-zenes **28** and bisarylmethyldicyanobenzenes **29** (**Scheme 3.5**) [23]. In the case of **25d/26**, 9-fluorenone (80%) and 9-fluorene **27d** (18%) were the major products. For **25a/26**, **25b/26** and **25c/26**, the corresponding arylmethanes **27** were not detected due to their volatile properties. In these solid-state photoreactions, the conversions of the

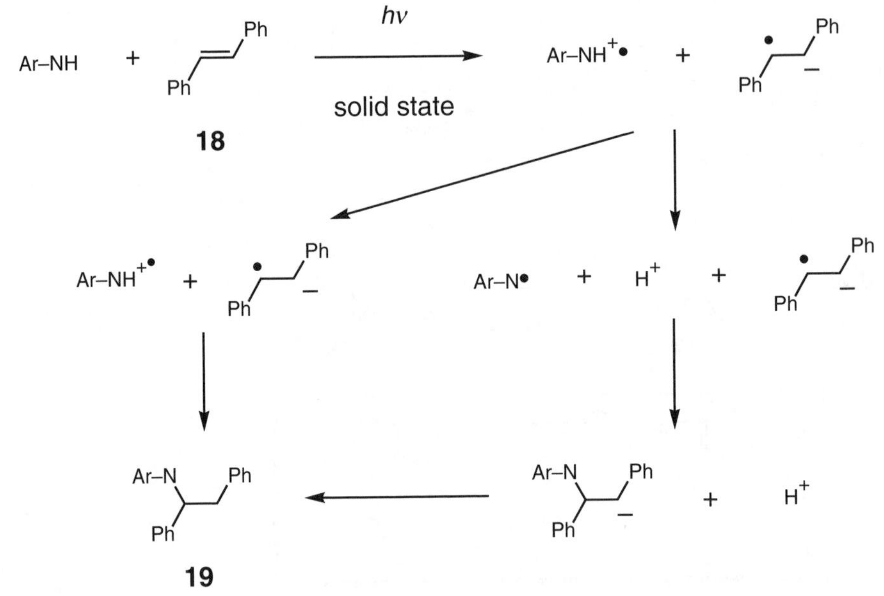

Scheme 3.4 Possible reaction mechanism of photoaddition of an NH compound to *trans*-stilbene.

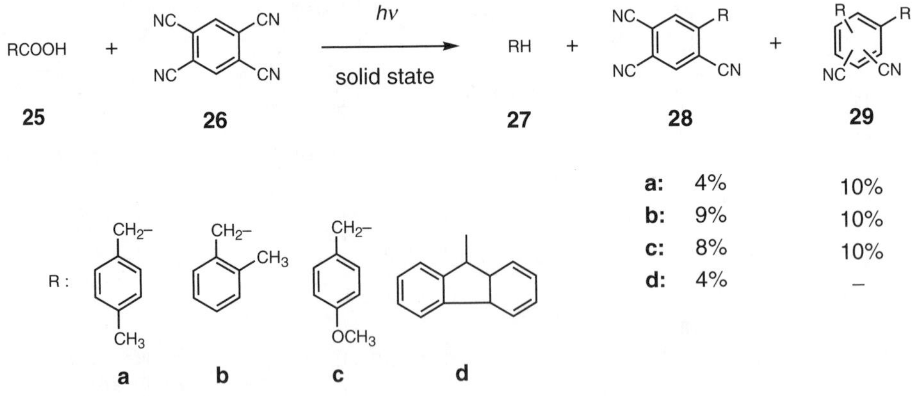

a:	4%	10%
b:	9%	10%
c:	8%	10%
d:	4%	–

Scheme 3.5 Decarboxylative condensation between arylacetic acids and TCNB.

starting acids **25** and TCNB were 87–100% and 13–44% respectively. Much more highly selective photodecarboxylation occurs in two-component crystals of arylacetic acids and acceptor molecules such as TCNB and aza aromatic compounds (see Section 8.1).

3.4. *[2 + 2] Photocycloaddition*

Simple polycrystalline mixtures obtained by melting and resolidification of various heterocyclic and aromatic aldehydes **30a–g** and excess indole **12** undergo a solid-state photoreaction to give condensation products **31a–g** (**Scheme 3.6**) [24,25]. The molecular structures of **31a**, **31c** and **31d** were confirmed by X-ray crystallographic analysis. The major product from the crystalline mixture of **30d/12** was a tetraindolyl compound **33** (25%), in addition to **31d** (12%) and an unstable unknown product (8%).

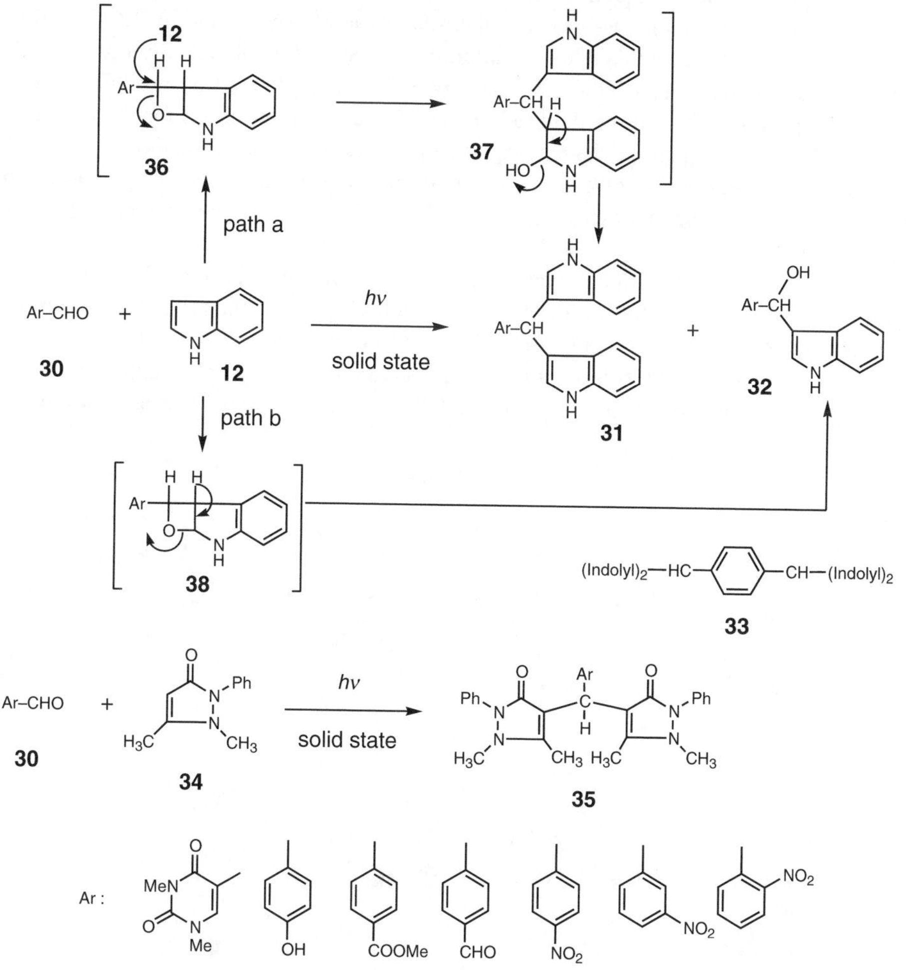

Ar :	MeN O N Me	OH	COOMe	CHO	NO$_2$	NO$_2$	NO$_2$
Yield (%)	**a**	**b**	**c**	**d**	**e**	**f**	**g**
31	40	15	45	12, 25*	25	20	36
32	–	–	–	–	17	28	34
35	61	11	–	–	–	–	–

* Yield of **33**

Scheme 3.6 Photocondensation via [2 + 2] cycloaddition.

Similar solid-state photocondensation occurred between aldehydes **30** and antipyrine **34**. Irradiation of the simple crystalline mixture of **30b/34** gave **35b** (11%) and *p*-hydroxybenzoic acid (38%). 5-Formyl-1,3-dimethyluracil **30a** and **34** form a molecular compound, which gave **35a** on irradiation.

This type of solid-state condensation has been found to show a substituent effect. Solid mixtures between *p*-, *m*- or *o*-nitrobenzaldehyde **30e–g** and indole **12** gave nitrophenyl-(3-indolyl)-methanols **32e–g** in addition to **31e–g**. A dual pathway involving the oxetane intermediate **36**, which is formed by a [2 + 2] cycloaddition product between the C = O group of **30** and the double bond of **12**, has been proposed for the formation of **31** and **32** [26]. In path (a) **36** undergoes ring cleavage to react with another molecule of indole **12** to eventually give **31**, whereas in path (b) **38** undergoes ring cleavage to give directly **32**. In the latter case, the process from **36** to **37** may be destabilized by the nitro group.

Scheme 3.7 shows additional examples of the [2 + 2] cycloaddition. Upon solid-state irradiation of a 1:6 crystalline mixture of 1-methyl-2,4,5-triphenylimidazole **39** and benzophenone **1**, a stable oxetane **40** is obtained in moderate yield [27]. The 1:2 crystal prepared by melting and resolidification of uracil derivative **41** and phenanthrene **13** gave selectively a *cis*-adduct **43** in 98% yield [28]. The 1:2 crystal of **41** and acenaphthylene

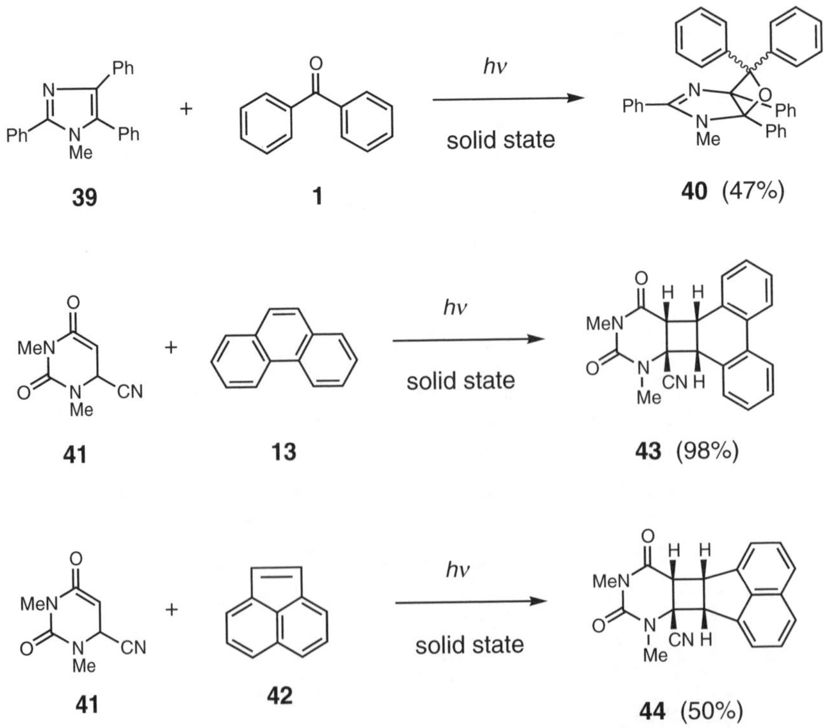

Scheme 3.7 [2 + 2] Photocycloaddition in simple polycrystalline mixtures.

42 afforded also *cis*-adduct **44** as the major product. Two stereoisomeric photodimers of **42** are also obtained as minor products.

3.5. Control factors for photoreactions occurring at the polycrystalline interface

3.5.1. Proximity effect

In the solid-state photoreactions of simple polycrystalline mixtures, the close contact of the crystal surfaces at the interfaces of the component crystallites is one of the important conditions for reaction. In order to determine the proximity effect in the solid-state photoreaction between aromatic aldehydes **30** and indole **12** (**Scheme 3.6**), a solution phase irradiation was carried out. Irradiation of a solution of 5–19 mM aromatic aldehydes **30a–d** and 38–51 mM indole in acetonitrile gave no condensation products **31a–d** [24,25]. However, when a liquid mixture of indole and excess benzaldehyde is irradiated, **31** (Ar≡Ph) is obtained in 75–85% yield.

3.5.2. Size effect of component crystallites

In the solid-state photoreaction of a simple mixture of microcrystals, close contact of the crystal surfaces at the interface of the component crystallites is one of the important conditions for reaction. It is essential therefore for the occurrence of bimolecular photoreactions at the polycrystalline interface of the two components that the crystalline mixture has as large a surface area as possible. A large surface area will also increase the conversion of the starting materials. For the photoaddition in **Scheme 3.3**, the 1:1 solid mixtures of carbazole **15** and *trans*-stilbene **18** were prepared by cooling at different temperatures of −78 °C, 20 °C, and 100 °C [20]. The PXD patterns show that they consist of a simple mixture of the crystallites of **15** and **18**, since the PXD patterns are just the sum of those of both component crystals (Figure 3.2). The crystallite size is defined as the minimum size of a single crystal structure. A single particle is assembled from a large number of crystallites (Figure 3.3). The crystallite sizes were determined by the Scherrer method based on the independent peaks of $2\theta = 9.16°$ and $2\theta = 12.06°$ for **15** and **18**, respectively (Figure 3.2). Table 3.1 shows the effect of the cooling temperature of the 1:1 melt of carbazole **15** and *trans*-stilbene **18** on the crystallite sizes and the

Table 3.1 Crystallite size in the 1:1 simple mixture of carbazole and *trans*stilbene and its photoreactivity

Cooling temperature (°C)	Crystallite size (Å)		Yield of adduct **19** after 3-h irradiation (%)
	Carbazole	*trans*-Stilbene	
−78	294	342	5.1
20	364	381	3.9
100	412	415	3.1

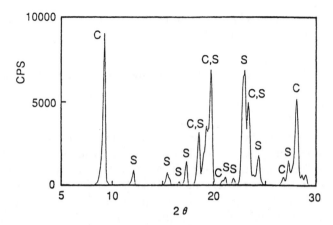

Figure 3.2 PXD patterns of the simple crystalline mixture of carbazole and *trans*-stilbene. C and S denote the peaks due to carbazole and *trans*-stilbene, respectively.

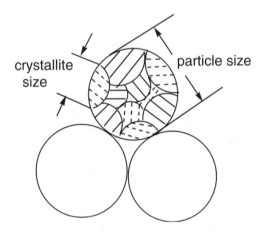

Figure 3.3 Particle size and crystallite size of a simple polycrystalline mixture of two components.

formation rate of the photoadduct **19**. Faster cooling of the melt gave smaller-sized crystallites of **15** and **18** and higher rates of adduct formation.

3.5.3. *Energy transfer in polycrystalline mixtures*

The exiton migration theory explains the fast energy transfer in the crystalline state [29]. Irradiation of anthracene crystals doped with a trace (10^{-4} mol) of naphthacene leads to the emission of naphthacene fluorescence only. The simple polycrystalline mixture of indole **12** and phenanthrene **13** prepared by melting and resolidifying emits only phenanthrene fluorescence, showing that a facile energy transfer occurs through the crystallite interface to form the fluorescent state of phenanthrene having a lower singlet level [21]. This indicates that the photochemical formation of adduct **14** in the solid state (**Scheme 3.3**) occurs from the singlet state of phenanthrene. On the other hand, the 1:1 crystal of indole and naphthalene, which consists of a molecular compound and forms a photoadduct of type **14** in the solid state, emits only indole fluorescence, showing a facile energy transfer in the crystal lattice of the molecular compound to the component having a lower single level.

The luminescence spectra of the simple crystalline mixtures between NH heteroaromatics (**12**, **15** and **16**) and *trans*-stilbene **18** (**Scheme 3.3**) were measured: mixture **12/18** emits only *trans*-stilbene fluorescence, while **15/18** and **16/18** emit carbazole **15** and phenothiazine **16** phosphorescence, respectively. These results indicate that the excitation energy migrates smoothly through the interfaces of the mixed microcrystals of both components to result in the formation of the fluorescent or phosphorescent state of the component having a lower singlet or triplet level.

The fluorescence spectra of polycrystalline mixtures of arylacetic acids **25a–d** and TCNB **26** (**Scheme 3.5**) were approximately the same as the sum of those of the two components, showing lower emission intensities. Only in the case of the solid mixtures **25a/26** and **15b/26** did new peaks appear, which were ascribed to excimer fluorescence at 457 and 428 nm, respectively. This fluorescence behavior is different from facile energy transfer through the interface of the polycrystalline mixtures of the above two cases.

3.5.4. *Other factors*

Some selectivities were observed for the solid-state photoreactions of most of the solid mixtures in comparison with their solution phase photoreactions. These have been described in Sections 3.1 to 3.4. However, in certain cases, a similar product ratio was observed for both the solid-state and solution-phase photoreactions. Due to the experimental difficulties in elucidating the exact molecular arrangement at the interface of polycrystalline mixtures, further experimental developments are required in the future to provide further information.

The other important problem is whether the solid-state photoreactions at the interface continue until there is total conversion of the substrates. In order to throw light on this problem, the quantum yield of the solid-state photoreaction of the simple polycrystalline mixtures of *p*-tolylacetic acid **25a** and TCNB **26** was measured using a thin film technique [30]. The quantum yields were 0.04 and 0.20 in the solid state and acetonitrile solution, respectively. The actual conversions of **25a** and **26** were 100% and 99% respectively after 4 h of irradiation of the crystals, showing that solid-state photoreactions at the interface proceed effectively in certain cases despite the low mobility of the molecules.

4. Hydrogen Abstraction

Hydrogen abstraction by the excited carbonyl group is the most typical photoreaction in both the solution phase and in the solid state. In the case of intramolecular hydrogen abstraction in the crystalline state, a large number of examples are already known and geometric requirements have been precisely discussed by Scheffer [3]. In contrast, inter-molecular hydrogen abstraction reactions occurring in two-component crystals are scarcely reported. Herein only two examples are presented, namely a molecular crystal of duroquinone and durene, and a series of inclusion complexes of deoxycholic acid with ketones.

4.1. Molecular crystal of duroquinone and durene

A yellow 2:1 two-component crystal **8•9** was prepared by crystallization of duroquinone **8** and durene **9** from ethyl acetate (**Scheme 4.1**) [18,31]. Melting and resolidification of the 2:1 mixture of both components also gave crystalline **8•9**. Figure 4.1 illustrates the phase diagram of the two components without solvent constructed from measurements of DSC and PXD patterns. The melting point of **8•9** is 95 °C, appearing between those of **8** (112 °C) and **9** (79 °C). X-ray crystallographic analysis revealed that the space

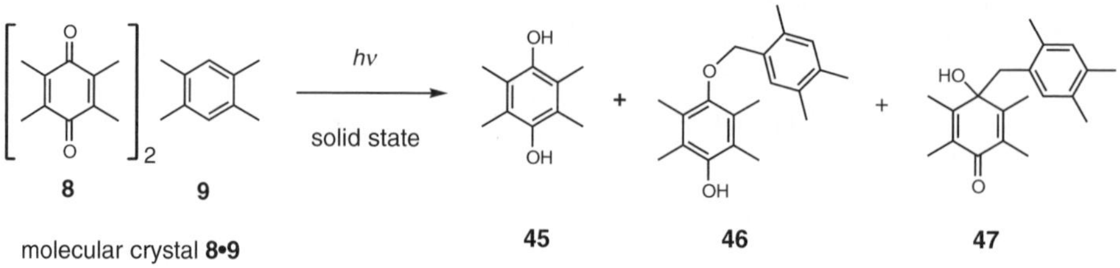

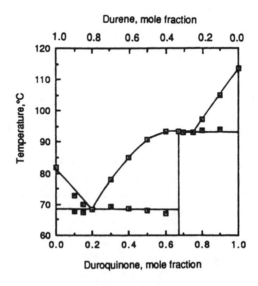

Scheme 4.1 Photoreaction in the two-component crystal of duroquinone and durene.

Figure 4.1 Phase diagram of duroquinone and durene. *Source*: from ref. 18 with permission. © 1994 Elsevier Science.

group is $P\bar{1}$ and the unit cell is quite large; $a = 16.106(7)\,\text{Å}$, $b = 17.580(9)\,\text{Å}$, $c = 18.759(7)\,\text{Å}$, $\alpha = 58.48(3)°$, $\beta = 63.49(3)°$, $\gamma = 65.83(3)°$, $D_c = 1.169\,\text{g cm}^{-3}$, $Z = 6$, $R = 0.11$ (Figure 4.2). There are 12 molecules of **8** and 6 molecules of **9** in each unit cell. The rings 3, 4, 5 and 8 are duroquinones, and the rings 6 and 7 are durenes. Rings 1 and 2 exhibit disorder in which both duroquinone and durene coexist at about 50% occupancy. The rings 7, 4, 2, 3 and 8 are nearly parallel with interplane distances of 3.4–3.6 Å. The other rings 6, 9, 1, 5 and 10 are parallel but lie perpendicular to the rings 7, 4, 2, 3 and 8. The interplane distances between 6 and 9 and also 5 and 10 are 3.5 Å. This layer structure suggests that **8•9** might be a CT complex. The CT interaction is very weak, however, because a significant CT band is not observed and the yellow color is almost the same as that of duroquinone **8**.

Irradiation of the pulverized crystals caused hydrogen abstraction and radical coupling to give durohydroquinone **45** and the two products **46** and **47** (**Scheme 4.1**). The product distributions were dependent on the irradiation wavelength (Table 4.1). UV

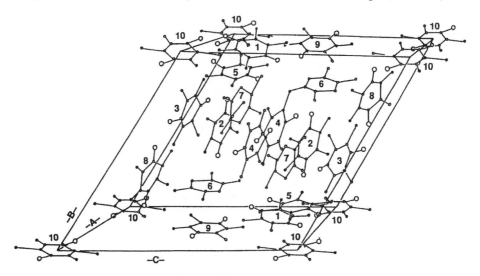

Figure 4.2 Molecular arrangement within the two-component crystal formed between duroquinone and durene. *Source*: from ref. 31 with permission. © 1994 Elsevier Science.

Table 4.1 Photoreaction of the two-component crystal of duroquinone and durene.

	Irradiation		Conversion (%)		Yield (%)		
Lamp	Wavelength (nm)	Temperature (°C)	**8**	**9**	**45**	**46**	**47**
Hg	>290	10	64	51	20	7	3
Xe	>290	10	48	46	4	3	16
Xe	>290	−50	31	44	–	–	15
Xe[a]	>390	10	42	41	3	3	20

[a] Through UV cut filter.
Source: from ref. 34 with permission. © 1987 Am. Chem. Soc.

irradiation (>290 nm) with a high-pressure mercury lamp afforded durohydroquinone **45** (20%) as the main product, while irradiation (290 nm and >390 nm) with a xenon short arc lamp gave mainly the adduct **47** (20%). In contrast, solution photolysis of **8** and **9** in acetonitrile or benzene gave different products from the solid-state photoreaction, but these could not be identified due to their instability, confirming that the reaction is specific in the crystalline state.

Scheme 4.2 shows the possible reaction mechanism. The excited quinone carbonyl group abstracts a hydrogen atom from an adjacent durene methyl group to give the intermediate duroquinone radical **48** and duryl radical **49**. An adduct **47** is produced by subsequent radical coupling of **48** and **49**. Hydroquinone **45** is formed by disproportionation of the semiquinone radical **48a** produced by rearrangement of **48**. An electron spin resonance (ESR) signal was observed on irradiation of the crystal **8•9** at −100 °C. This is thought to be due to **48a** from the g-value of 2.0051, the peak-to-peak width of 2.97 mT, and the hyperfine structure with peak to peak width of 0.69 mT (Figure 4.3).

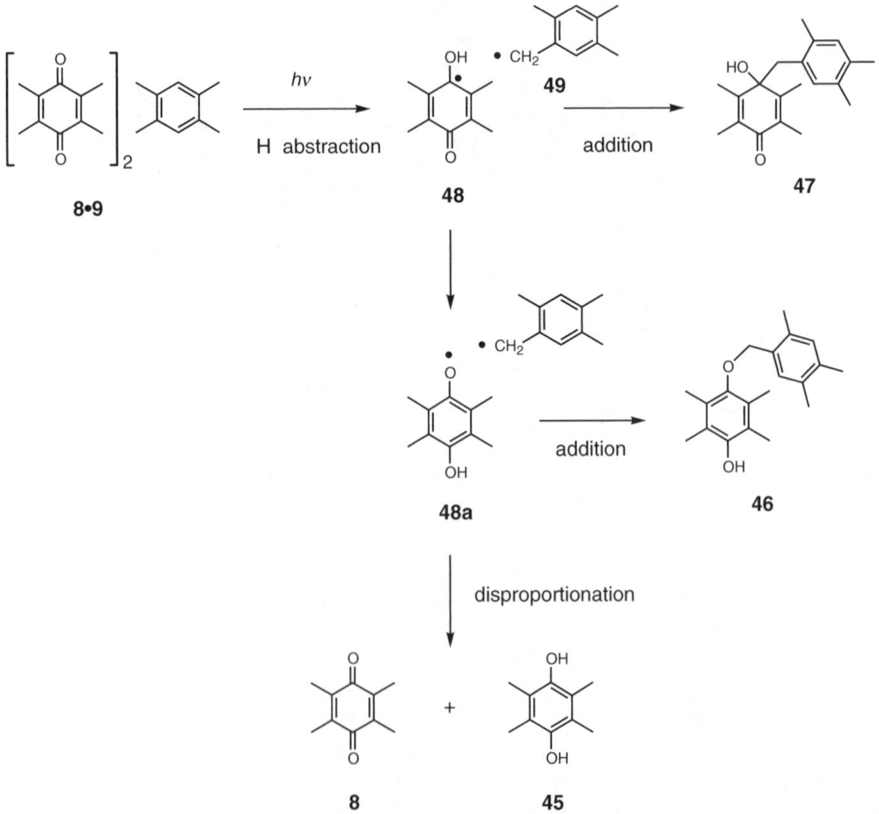

Scheme 4.2 Possible reaction mechanism for the photoreaction in a two-component crystal of duroquinone and durene.

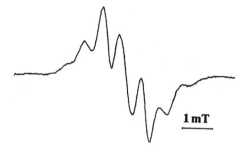

Figure 4.3 ESR spectrum of the two-component crystal from duroquinone and durene under UV irradiation at −100°C. *Source*: from ref. 31 with permission. © 1994 Elsevier Science.

The radical was relatively stable since about half of the signal strength remained after completion of irradiation and keeping the sample in the dark at 30°C for 1 h.

One of the important factors for intermolecular hydrogen abstraction is that the distances between the oxygen atoms of quinone and the hydrogen atoms of durene methyl group are short enough. This is estimated to be 2.2–2.7 Å by Scheffer [3]. Such hydrogen abstraction by **8** from **9** is possible in the lattice due to the layer stacking arrangement, which has a short interplane distance of 3.5 Å between the rings 9 (**8**) and 6 (**9**) and also between 10 (**8**) and 5 (**9**) (Figure 4.2). The C = O•••H₃C– distances are estimated to be in the range 2.2–2.8 Å from the crystal data, which are short enough to allow the quinone C = O to directly abstract the hydrogen to give the semiquinone radical **48a**.

4.2. Inclusion complexes of deoxycholic acid with ketones

It is well known that deoxycholic acid and apochlolic acid form inclusion complexes with a number of guest molecules. Intermolecular hydrogen abstraction of allylic hydrogens by ketones in the solid state occurs in the crystalline complexes of cholic acid with ketones such as acetone, acetophenones and propiophenones [32–35]. Such reaction processes have been precisely studied by X-ray crystallographic analysis.

The crystalline inclusion complexes **50•a**, **50•b** and **50•c** were obtained from solutions of deoxycholic acid **50** with acetophenone **a**, *m*-chloroacetophenone **b** and *p*-fluoroacetophenone **c** in methanol. The host:guest molar ratios are 5:2, 3:1 and 8:3, respectively (**Scheme 4.3**). The molecular arrangements in the three crystals are similar to those in other reported inclusion complexes of **50** (Figure 4.4). Guest molecules are included in the cavities between the steroid molecules and occupy the channel structures along the *c*-axis.

UV irradiation (<290 nm) of these crystals for about 30 days at room temperature gave the corresponding single diastereomeric photoadduct **51** (**Scheme 4.3**). A reasonable reaction process would be for the excited carbonyl oxygen of acetophenone to abstract the hydrogen atom H–C5 of **50** and then for the steroid carbon radical C5 and the sp³ carbon radical derived from the carbonyl carbon to couple.

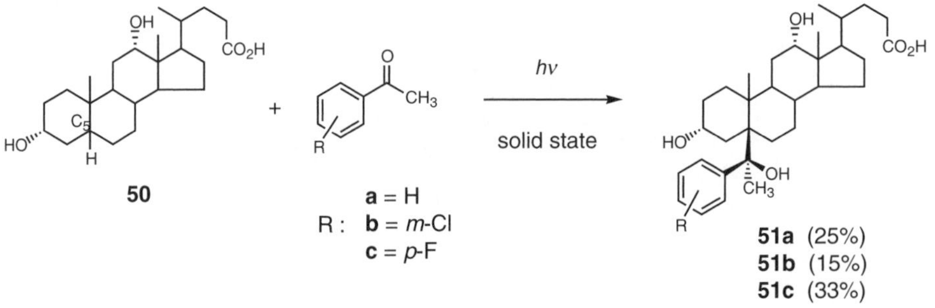

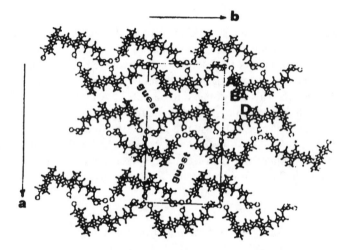

Scheme 4.3 Photoreaction in the inclusion crystals of deoxycholic acid and acetophenones.

Figure 4.4 Packing arrangement of deoxycholic acid molecules in the α motif viewed along the channel
c-axis. The guest molecules are not shown. *Source*: from ref. 33 with permission. © 1985 Am. Chem. Soc.

However, an important problem remained, since the newly generated chiral carbon
had the (*S*)-configuration, which is opposite to that expected from the molecular
arrangement. This indicates that photoaddition of the guest molecule to C5 takes place
with a net rotation of 180° of the guest acetyl group, as shown in **Scheme 4.4**.

X-ray crystallographic analysis of two complexes **50·a** and **50·c**, before and after irradiation, were performed at low temperature (−170 °C). Fortunately these reactions were
revealed to proceed via single-crystal-to-single-crystal transformation without decomposition of the initial crystal structures [34]. The X-ray crystal data for **50·a** before and after
irradiation are summarized in Table 4.2. The space group $P2_12_12_1$ is unchanged and the cell
constants are almost coincident before and after irradiation. Furthermore, the reaction pathway was successfully traced by X-ray analyses of crystals irradiated for different times up
to 67 days. Figure 4.5 shows the stereoviews obtained at various stages of the reaction
process. These reveal that the ketyl group rotated around the C(ketyl)–C5(phenyl) bond
(b and c) and finally coupled with the steroid C5 position to produce the adduct (d).

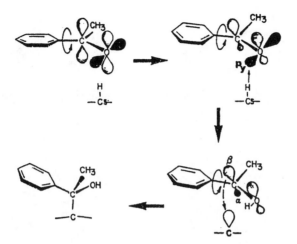

Scheme 4.4   Photoreaction process accompanied by the rotation of the acetyl group in the inclusion crystal of deoxycholic acid and acetophenone. *Source*: from ref. 34 with permission. © 1987 Am. Chem. Soc.

Table 4.2   Crystal data of the complex **50•a** of deoxycholic acid **50** and acetophenone **a** at −170 °C before and after irradiation

	Before irradiation	After irradiation
⎰ formula	$C_{24}H_{40}O_4 \cdot 2/5(C_8H_8O)$	
⎱ stoichiometry (occupancy)		
deoxycholic acid **50**		⎰ $C_{24}H_{40}O_4$
		⎱ 0.871(8)
photoproduct **51a**[a]		⎰ $C_{32}H_{48}O_5$
		⎱ 0.129(8)
acetophenone **a**		⎰ C_8H_8O
		⎱ 0.232(17)
space group	$P2_12_12_1$	$P2_12_12_1$
a (Å)	25.243(7)	25.173(5)
b (Å)	13.606(2)	13.685(3)
c (Å)	7.198(2)	7.178(4)
V (Å^3)	2475	2472
Z	4	4
D_{cac}	1.17	1.17
R	0.038	–
crystal size (mm)	$0.2 \times 0.3 \times 0.6$	$0.2 \times 0.3 \times 0.6$

[a] Sum of occupancies of deoxycholic acid + product = 1.

Furthermore, the stereoviews demonstrated that during the course of photoconversion there is minimal motion of the phenyl ring of acetophenone. Hence, in the case of single-crystal-to-single-crystal reactions, it becomes possible to elucidate the reaction path through X-ray crystallographic analysis, which is a unique advantage of solid-state reactions.

207

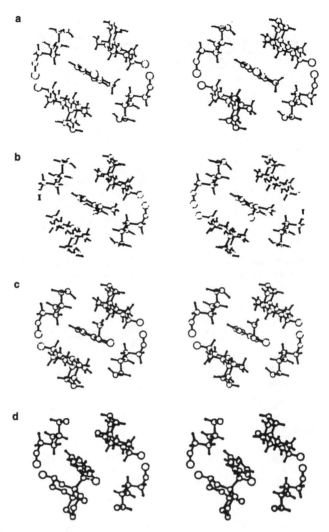

Figure 4.5 Stereoviews of the minimized positions of the ketyl radical and part of the surrounding neighboring deoxycholic acid moieties forming the channel wall as the ketyl group rotates about the C(ketyl)–C(phenyl) bond. The first (a) corresponds to the unreacted crystal and the final (d) to the photoproduct **51a** in the crystal of deoxycholic acid with acetophenone. *Source*: from ref. 34 with permission. © 1987 Am. Chem. Soc.

5. Addition

Addition reaction is most important in organic synthesis. Four types of photoaddition reactions between different molecules in two-component crystals are presented here, while [2 + 2] photocycloaddition will be discussed later. An important advantage in utilizing a two-component crystal as a reactant is that new reactivity via photoinduced electron transfer can be induced by combining electron donor and acceptor species in the two-component crystal. An interesting example shows that two kinds of crystals formed from acridine and phenothiazine have quite different photoreactivities despite the occurrence of similar photoinduced electron transfer. Addition reactions in a CT crystal of TCNB and benzyl cyanide [38], and in two-component crystals of NH heteroaromatic compounds with aromatic compounds [19,39] are also included. A new type of photopolymerization is shown by a series of ammonium salts of 1,3-diene mono- and dicarboxylic acids, which is different from the well-known [2 + 2] polymerization in the solid state [40–42].

5.1. Photoaddition of two kinds of crystals from acridine and phenothiazine via electron transfer process

An interesting example was discovered where yellow crystals **Y52•16** and red crystals **R52•16** are simultaneously formed by crystallization from a solution of acridine **52** and phenothiazine **16** in acetonitrile. The yellow crystal gives an adduct on irradiation, while the red crystal has almost no photoreactivity (**Scheme 5.1**) [36]. Further, the first evidence for the occurrence of solid-state photoinduced electron transfer was obtained by the successful measurement of transient absorption spectra in femtosecond diffuse reflectance spectroscopy.

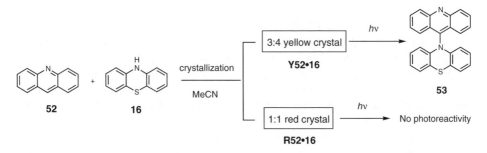

Scheme 5.1 Photoreaction of yellow crystals and red crystals formed from acridine and phenothiazine.

The yellow crystal **Y52•16** has a 3:4 molar ratio of acridine **52** and phenothiazine **16**, and the red crystal **R52•16** 1:1. By seeding, the desired crystals can be easily prepared on a large scale. Although the yellow and red colors of the single crystals are visually very different, the two spectra seem to be quite similar at longer wavelengths. The melting point of the yellow crystals is 134 °C, falling between those of **52** (107 °C) and **16** (185 °C). The red crystal undergoes phase transition to the yellow crystal at around 90 °C, and then melts at 134 °C.

X-ray crystallographic analysis revealed that the structures of the two crystals were quite different. The yellow crystal **Y52•16** belongs to space group *C2/c*, and the cell parameters are $a = 30.194(3)$ Å, $b = 14.686(3)$ Å, $c = 15.996(3)$ Å, $\beta = 107.930(9)°$, $Z = 4$, $D_c = 1.314$ g cm^{-3}, $R = 0.089$. The molecular arrangement is relatively uncommon, with twelve and sixteen molecules of acridine and phenothiazine, respectively, packed in the unit cell (Figure 5.1a). The molecular shape of acridine is almost flat, but

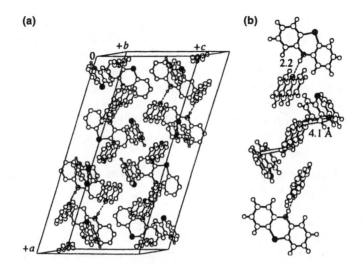

Figure 5.1   (a) Molecular arrangement in the yellow crystal and (b) the radical coupling path. In (a) the hydrogen atoms are omitted except for the NH hydrogens for clarity. Solid circle, sulfur atom; cross circle, nitrogen atom. Dotted atoms in the disordered acridine molecules represent the coexistence of carbon and nitrogen atoms with occupancies of 50%. *Source*: from ref. 36 with permission. © 1997 The Royal Society of Chemistry.

that of phenothiazine is butterfly-like. A hydrogen-bonded pair of acridine and phenothiazine is formed through the N•••H–N bonding (2.15 Å); the plane of **52** and the central C–C–C–C plane of **16** are almost perpendicular (86.58°). Eight such hydrogen-bonded pairs are packed in a head-to-tail manner with a distance of 6.4 Å between the two acridine planes. The eight molecules of acridine and phenothiazine are arranged almost parallel to the *ab* face and the *ac* face, respectively. The residual four acridine molecules are disordered, in an arrangement where N and C atoms coexist at about 50% occupancy (Figure 5.1). The four disordered acridine molecules do not form hydrogen bonds, and are situated between the two phenothiazine molecules at a dihedral angle of 58.2°. Furthermore, the acridine molecules with and without hydrogen bonds are arranged almost parallel to each other (dihedral angle 5.15°). On the other hand, the phenothiazine molecules with and without hydrogen bonds are almost perpendicular (dihedral angle 89.38°).

The red crystal **R52•16** has space group $P2_1/c$, and the cell parameters are $a = 13.121(2)$ Å, $b = 9.12(3)$ Å, $c = 17.318(2)$ Å, $\beta = 110.83(1)°$, $Z = 4$, $D_c = 1.297$ g cm^{-3}, $R = 0.056$. In the crystal lattice, a similar hydrogen-bonded pair of **52** and **16** is formed with an H•••N distance of 2.19 Å; the plane of **52** and the central C–C–C–C plane of **16** are almost perpendicular (85.73°) (Figure 5.2). Four molecular pairs are arranged in a head-to-tail manner in the unit cell to form two independent stacking columns (A) and (B) with interplanar distances of 3.9 and 8.9 Å between the two acridine molecules. The phenothiazine molecules in column (A) are arranged almost perpendicular to the acridine planes in the next stacking column (B), with the shortest distance of 2.9 Å between the aromatic H atom of phenothiazine and the acridine plane. Such rigid packing is a feature of the red crystal.

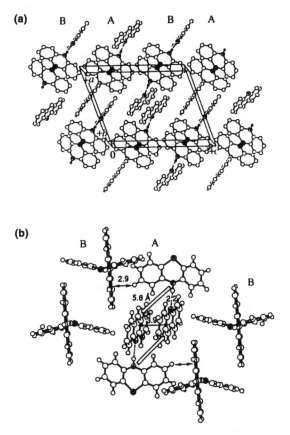

Figure 5.2 (a) Molecular arrangement in the red crystal and (b) the radical coupling path. In (a) the hydrogen atoms are omitted except for the NH hydrogens for clarity. Solid circle, sulfur atom; cross circle, nitrogen atom. (A) and (B) are the independent molecular stacking columns. *Source*: from ref. 36 with permission. © 1997 The Royal Society of Chemistry.

The yellow crystals of **Y52•16** were pulverized in a mortar and irradiated at >290 nm with a 400-W high-pressure mercury lamp through Pyrex glass under argon. The UV irradiation caused an addition reaction, followed by dehydrogenation, to give **53** in 14% yield as the sole product (**Scheme 5.1**). In contrast, the red crystal **R52•16** showed much lower photoreactivity and afforded no photoproduct **53**. This difference must be attributed to the difference in crystal structures because both consist of the same components. For a comparison, photolysis of a mixture of **52** and **16** in acetonitrile solution resulted in similar photoaddition to give **53** and biacridane in 64% and 35% yield, respectively.

Figure 5.3 shows the transient absorption spectra of the yellow microcrystals of **Y52•16** measured by femtosecond diffuse reflectance spectroscopy, which is a new technique [37]. The absorption bands with maximum wavelengths at around 600 and 520 nm were observed immediately after excitation with a 390-nm laser pulse. They were assigned to the absorption spectrum of acridine anion radical and that of the phenothiazine cation radical, respectively, on the basis of the similarity to these in solution. This

211

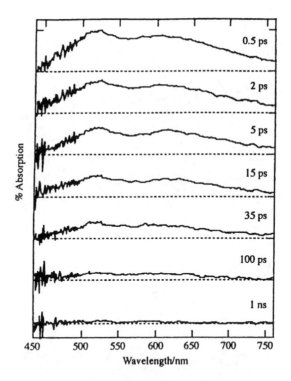

Figure 5.3 Transient absorption spectra of the yellow crystal from acridine and phenothiazine excited at 389 nm. *Source*: from ref. 36 with permission. © 1997 The Royal Society of Chemistry.

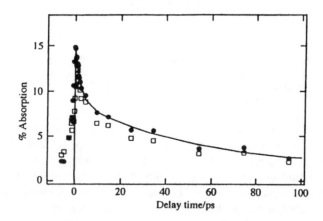

Figure 5.4 Observed time profiles of the transient absorption of the yellow crystal. Probe wavelengths are at 524 nm (black circle) and 613 nm (square). *Source*: from ref. 36 with permission. © 1997 The Royal Society of Chemistry.

provides the first direct evidence for the generation of ionic species in a photoreactive two-component crystal. The decay of transient absorption in Figure 5.4 is related to non-exponential behavior, and it can be approximated to a double-exponential function with lifetimes of 2 ps and about 50 ps, reflecting the uncommon molecular arrangement of coexisting hydrogen-bonded pairs and non-hydrogen-bonded pairs in the lattice.

In the case of red microcrystals of **R52•16**, similar transient absorption spectra to those of the yellow crystals were observed, but the decay approximated to a single exponential decay with a lifetime of about 2 ps and no transient absorption was observed after several tens of picoseconds. The fast decay components observed for the two crystals are probably due to the CT excited state of the hydrogen bond pairs. Compared with the CT excited state of some typical crystals with weak CT complexes, this lifetime is very short, suggesting that hydrogen bonding is the key role in the ultrafast deactivation of the excited state. In contrast, a slow deactivation observed only for the yellow crystals corresponds to a charge-separated species formed between the non-hydrogen-bonded pairs. If we consider that a part of them results in the formation of photoproduct **53**, then the above results for transient absorption measurements are quite consistent with the different photoreactivities of the yellow and red crystals.

Scheme 5.2 shows the possible reaction mechanism based on the transient absorption spectra and the crystal sructure. Upon irradiation, photoinduced electron transfer from **16** to **52** gives the acridine anion radical and the phenothiazine cation radical, followed by proton transfer to produce the hydroacridine radical **54** and the dehydrophenothiazine radical **55**. In the case of crystalline **Y52•16** (Figure 5.1b), radical coupling between **54** and **55**, which originates from the disordered acridine and non-hydrogen bonding phenothiazine, can occur most effectively over the shortest distance of 4.14 Å and is followed by dehydrogenation to give the product **53**. The occurrence of radical coupling between the hydrogen bonding pair is difficult because a half rotation of **54** is needed in the crystal lattice. The low yield of the product **53** is most probably due to the existence of only four disordered acridine molecules out of the twelve in the unit cell.

On the other hand, the lack of product **53** on irradiation of the red crystal should be attributed not only to the short lifetimes of the radical species but also to the molecular arrangement in the crystal lattice. The distance of 5.58 Å for the radical coupling of **54**

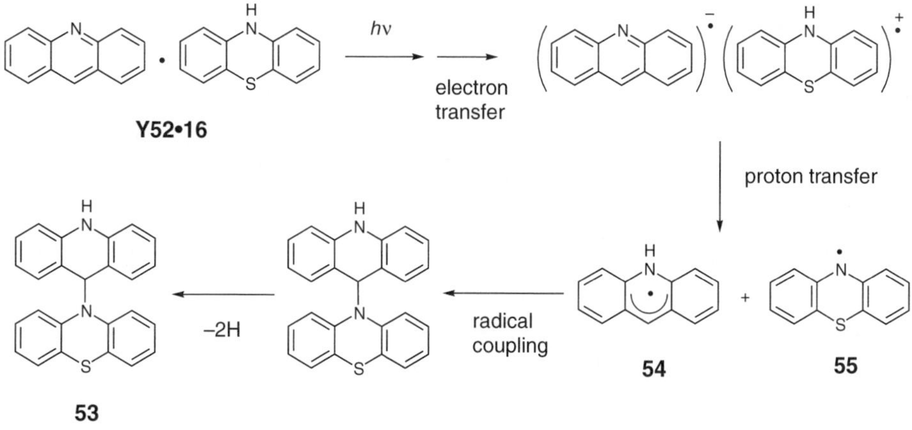

Scheme 5.2   Possible reaction mechanism in the yellow crystal of acridine and phenothiazine.

and **55** in Figure 5.2b is thought to be short enough for radical coupling, because an even longer distance of 6.8 Å results in radical coupling in other two-component crystals. Steric hindrance may also be a reason for the non-occurrence of radical coupling. In Figure 5.2b, the interplanar distance between two acridine molecules in column (A) is 3.9 Å, while the phenothiazine molecule of the stacking column (A) is close to the acridine molecule of the next stacking column (B), with a shortest distance of 2.9 Å between the H atom of phenothiazine and the acridine plane. Therefore, radical **55** cannot move to pair with radical **54** and this leads to no reaction in the lattice. The observation that such a rigid molecular packing arangement is not suitable indicates that some free space is required for the occurrence of crystalline state reaction.

5.2. A CT crystal of 1,2,4,5-tetracyanobenzene and benzyl cyanide

A 1:2 weak CT crystal **26•56** is obtained as colorless plates by crystallization of 1,2,4,5-tetracyanobenzene (TCNB) **26** (as an acceptor) from hot benzyl cyanide (BzCN) **56** (weak donor) (**Scheme 5.3**) [38]. The crystal is colorless and the CT transition band is observed around 367 nm. The crystal structure of **26•56** was determined by X-ray crystallographic analysis; the crystal data are triclinic, $P\bar{1}$, $a = 7.767$, $b = 10.957$, $c = 7.682$ Å, $\alpha = 93.80°$, $\beta = 117.61°$, $\gamma = 103.05°$, $Z = 1$, $D_c = 1.238 \, \mathrm{g\,cm^{-3}}$, $R = 0.053$. Two molecules of BzCN and one molecule of TCNB are arranged in a sandwich manner in the unit cell to form a charge transfer layer structure (Figure 5.5).

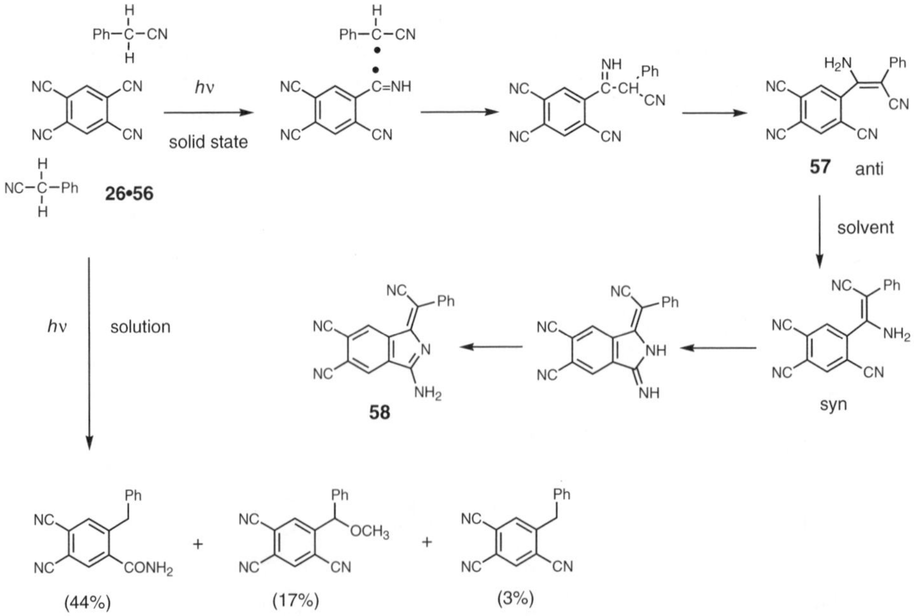

Scheme 5.3 Photoreactions in the CT crystal and in the solution of TCNB and benzyl cyanide.

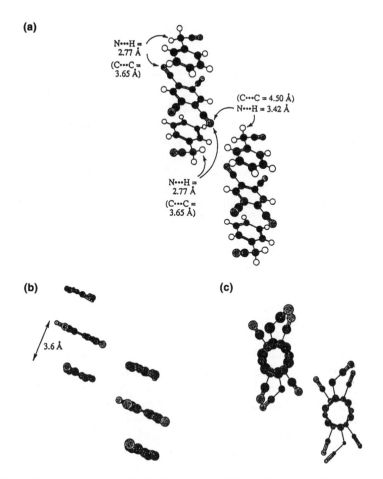

Figure 5.5 (a) Molecular arrangement in the CT crystal of TCNB and benzyl cyanide. (b) Parallel stacking horizontally rotated by −34°. (c) Superimposing horizontally rotated by 64°. For b and c, hydrogens are omitted for clarity. *Source*: from ref. 38 with permission. © 1997 Am. Chem. Soc.

Irradiation of the CT crystals **26•56** caused a coupling reaction giving the adduct **57** (**Scheme 5.3**). However, the product **57** easily cyclized in solution and formed an iso-indole derivative **58**, the structure of which was confirmed by X-ray crystallographic analysis. The consecutive reaction of **26•56** to **58** via **57** proceeded almost quantitatively at the 65% conversion of TCNB.

The possible reaction mechanism is shown in **Scheme 5.3**. Irradiation of crystals of **26•56** through a Pyrex filter (>290 nm) or a solution filter (BiCl$_3$ in dilute HCl, >355 nm) produced **57**, suggesting that the reaction is induced by CT excitation. Hydrogen abstraction by the cyano group of TCNB occurs next from the benzylic hydrogen of BzCN followed by radical coupling and then a 1,3-proton shift to give **57**. The initial product **57** is stable in the crystalline state due to the fixation of the amino group and the relevant *o*-cyano group in *anti*-relationship. However, in the solution phase, the confor-mation of **57** easily changes from *anti* to *syn* followed by cyclization to **58**.

215

Figure 5.5a shows two types of TCNB/BzCN close contact in the crystal lattice. The shortest distance between one of the cyano nitrogens of TCNB and the nearest benzylic hydrogen of BzCN is 2.77 Å, and the corresponding C—C distance is 3.65 Å. These are probably short enough for hydrogen abstraction by the cyano group followed by coupling of the resultant radical pair, because they are nearly comparable to the sum (2.75 Å) of the van der Waals radii for N and H and that (3.4 Å) of C and C.

Furthermore, the distance between the cyano nitrogen of TCNB and the second-nearest benzylic hydrogen of BzCN is 3.42 Å, and the corresponding C—C distance is 4.50 Å. Despite these being slightly longer, reaction possibly occurs because less molecular rotation is required during the reaction pathway from reactant **26•56** to the product **57**.

On the other hand, photoproducts formed in the solution phase (MeCN) of TCNB and BzCN were very different. Compounds **57** and **58** were not produced, but three other products were obtained instead (**Scheme 5.3**). It appears that the reaction was also initiated by photochemical hydrogen abstraction by the cyano group, in the same manner to the solid-state photoreaction. The geometries of hydrogen abstraction and the subsequent reaction courses are, however, very different between the crystalline-state and the solution-state processes.

5.3. NH heteroaromatic compounds and aromatic compounds

A 1:1 molecular compound **15•59a** was obtained by the melting and resolidifying process of carbazole **15** and anthracene **59a** [39]. The formation of molecular crystals was confirmed by the different X-ray diffraction pattern from those of either component and a new sharp melting point (223.5 °C) recorded by differential scanning calorimetry, which lies between those of anthracene (216.8 °C) and carbazole (245.6 °C). Irradiation of the crystal gave an addition product with a C–N bond and a dehydrogenated product in 7% and 13% yield, respectively (**Scheme 5.4**). Similarly, indole **12** and naphthalene

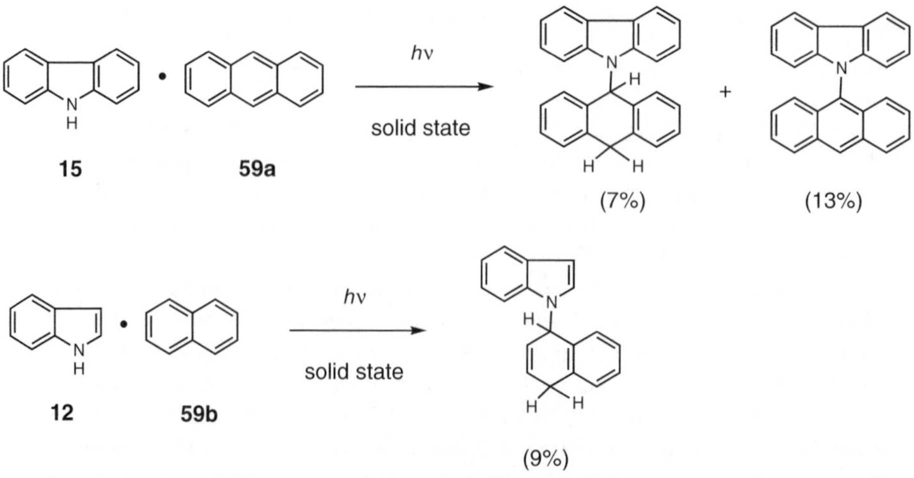

Scheme 5.4 Photoaddition in the two-component crystals of NH compounds and aromatic compounds.

59b also formed a two-component crystal by melting and resolidification of a 1:1 mixture of both components, and its solid-state photoreaction also yielded a photoadduct [19]. However, the X-ray structures of these two crystals were not determined due to the difficulty in obtaining high-quality single crystals. Other combinations of carbazole with aromatic hydrocarbons, such as naphthalene and phenanthrene, did not result in formation of molecular compounds. The polycrystalline mixtures of both components had similar photoreactivity (see Section 3.2).

5.4. *Photopolymerization in the ammonium salt crystals of 1,3-diene carboxylic acids*

Despite the fact that [2 + 2] cycloaddition of *trans*-cinnamic acids and [2 + 2] polymerization of 2,5-distyrylpyrazines are typical photoreactions in the solid state, a new type of photopolymerization was found in a series of ammonium salts of 1,3-diene mono- and dicarboxylic acids. The ammonium salt crystals incorporate muconic acid derivatives such as (*Z,Z*)-muconic acid **60**, (*E,E*)-muconic acid **61**, (*E,E*)-muconic acid monomethyl ester **62**, and (*E,E*)-sorbic acid **63**, and several amines such as benzylamine **a**, 1-naphthylmethylamine **b**, and the alkylamines **c** (**Scheme 5.5**) [40–42]. These

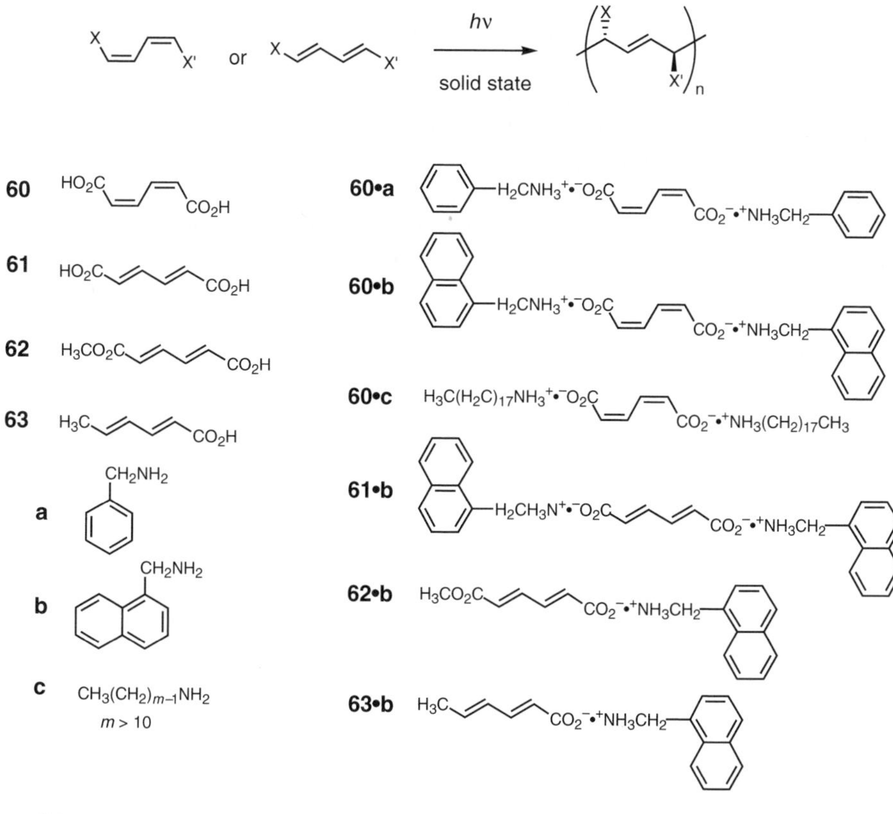

Scheme 5.5   Photopolymerization in the ammonium salt crystals with 1,3-diene carboxylic acids.

217

crystals were prepared by mixing the acids and amines in 1:1 ratios and mostly isolated as crystalline materials with higher melting points from appropriate solvents. Irradiation of the crystals with a high-pressure mercury lamp at room temperature under atmospheric conditions caused topochemical polymerization to give the corresponding polymers of the acids in high yield as the sole product (**Scheme 5.5** and Table 5.1). The reactant (Z,Z)- and (E,E)-isomers of acids were completely recovered without any isomerization.

X-ray crystal structure analysis revealed that column structures were formed between the acid and amine molecules through quaternary ammonium salt bridges and hydrogen bonding. Figure 5.6 illustrates the molecular arrangements in **60•a** as a representative example. The muconate anions are arranged in 1D fashion to yield a column structure. The primary ammonium cations act as triple hydrogen bond donors due to the three hydrogen atoms attached to the nitrogen atom, and the carboxylate anions act as triple or quadruple hydrogen bond acceptors due to the four lone-pairs of electrons. The polarized hydrogen network between them yields a one-dimensional columnar array. The other polymerizable monomer crystals have basically similar column structures despite the different d (layer-to-layer distance) values.

As shown in Figure 5.7, the diene moieties in the columns are arranged in a face-to-face manner and the dotted lines designate the carbons that make a new bond during polymerization. The tilt stacking of the muconates in the columns enables its double bonds to contact individually with the double bonds of different neighboring muconates along the column (Figure 5.6). The intermolecular carbon-to-carbon distance d_{cc} between the double bonds that topochemically react is 4.19 Å for **60•a**, which is sufficiently short for topochemical polymerization. However, the distance $d_{cc} = 5.37$ Å for **63•b** seems to be slightly larger than ideal for smooth polymerization. This is apparent from the difference in the structures of the sorbate and muconates in the crystals. Both ends of the muconate dianions are linked to the hydrogen-bond networks and sandwiched by the alkylammonium cation layers. In contrast, the monofunctional sorbate anions are attached to the hydrogen-bond network only at one side of the molecules. Therefore the conformation in the diene moiety of **63•b** readily changes when it reacts to make new bonds.

Table 5.1 Photopolymerization of ammonium muconates and sorbates in the crystalline state[a]

Run	Monomer	Polymer yield (%)	Isomer composition of monomer recovered		
			(Z,Z)-	(E,Z)-	(E,E)-
1	**60•a**	21	100	0	0
2	**60•b**	90	100	0	0
3	**60•c**	95	100	0	0
4	**61•b**	71	0	0	100
5	**62•b**	59	0	0	100
6	**63•b**	98	0	0	100

[a] Irradiated with a high-pressure mercury lamp for 8 h at room temperature.

(a)

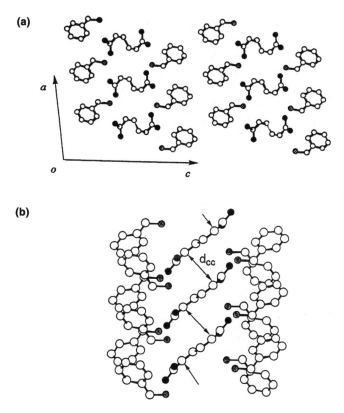

(b)

Figure 5.6 Molecular stacking in the column of **60•a**: (a) top view along *b*-axis and (b) front view along *c*-axis. Open, shadowed, and solid circles represent carbon, nitrogen, and oxygen atoms, respectively. Hydrogen atoms are omitted for clarity. *Source*: from ref. 40 with permission. © 1999 Am. Chem. Soc.

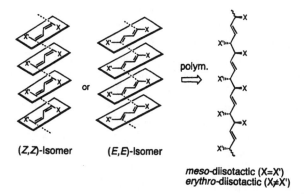

(*Z,Z*)-Isomer   (*E,E*)-Isomer

polym.

meso-diisotactic (X=X')
erythro-diisotactic (X≠X')

Figure 5.7 Translational molecular packing in the column of monomers and the stereochemical structure of polymers produced by topochemical polymerization of 1,4-disubstituted butadienes. *Source*: from ref. 41 with permission. © 2000 Am. Chem. Soc.

Furthermore, in the case of (Z,Z)-muconates **60•c** with N-alkylammonium, an interesting behavior was observed. When long-chain n-alkyl groups (carbon number $m = 10$–18) were introduced in the ammonium muconates, layered crystals were obtained with d values of 30–45 Å, which increased in proportion to the number of carbons in the N-alkyl group. The solid-state photoreaction gave insoluble and stereoregular polymer crystals, almost retaining the initial d values (28–44 Å). Next, the polymer crystals were dispersed in acidic methanol at room temperature and stirred for 30 min to convert them to poly(muconic acid) crystals. The d values were now radically changed to small 4.8 Å due to the removal of the long chain alkylamines. The isolated acid polymer was reversibly transformed into the ammonium polymer by reaction with the corresponding amine. Strikingly, a visible change in crystal volume was observed during the transformation. These transformations between the ammonium polymers and the acid polymers were completely reversible despite involving heterogeneous reaction.

6. [2 + 2] Cycloaddition

[2 + 2] Photocycloaddition has been known as a typical photoreaction in the crystalline state since 1964 [1]. A large number of [2 + 2] cycloadditions and polymerizations have been already reported, but most of them are reactions in one-component crystals. In contrast, [2 + 2] reactions in two-component crystals are scarcely known. Four examples are described herein [45–47,49], particularly stressing the correlation between the topochemical reaction paths and the geometrical molecular arrangements in these crystals.

6.1. *[2 + 2] Polymerization in a two-component crystal from different diolefinic compounds*

It is known that irradiation of crystalline 2,5-distyrylpyrazine **64** causes [2 + 2] cycloaddition and polymerization to give polymer poly-**64** (**Scheme 6.1**) [43]. Similarly ethyl 4-[2-(2-pyrazinyl)ethenyl]cinnamate **65** produces poly-**65** by the solid-state photoreaction [44]. A yellow 1:2 molecular crystal **64•65** could be prepared by crystallization from the solution of both components in benzene. The same crystalline material was also obtained by grinding **64** and **65** in a 1:2 ratio without solvent. The melting point of this substance is 166 °C, appearing between those of **64** (223 °C) and **65** (156 °C). Irradiation of the 1:2 crystal gave a 1:2 mixture of poly-**64** and poly-**65** (**Scheme 6.1**) [45]. The cross-polymer of **64** and **65** was not produced.

X-ray crystallographic analysis of the two-component crystal **64•65** was performed (Figure 6.1a). The crystal data are monoclinic, space group $P2_1/c$, $a = 30.380(5)$ Å, $b = 9.629(2)$ Å, $c = 7.494(1)$ Å, $\beta = 91.28(1)°$, $V = 2191.6(7)$ Å^3, and $Z = 4$, $D_c = 1.25$ g cm^{-3}, and $R = 0.088$. In the crystal, both molecules of **64** and **65** form separate columns along the c-axis with sheets perpendicular to the a-axis related by a glide mirror plane, indicating that the crystal is not a mixed crystal (solid solution) with a disordered molecular arrangement, but rather is a new molecular complex with an ordered molecular arrangement. The distances between the neighboring carbon–carbon double bonds in the same layer are 3.93 Å in the column of **64**, and 4.01 and 3.98 Å in the

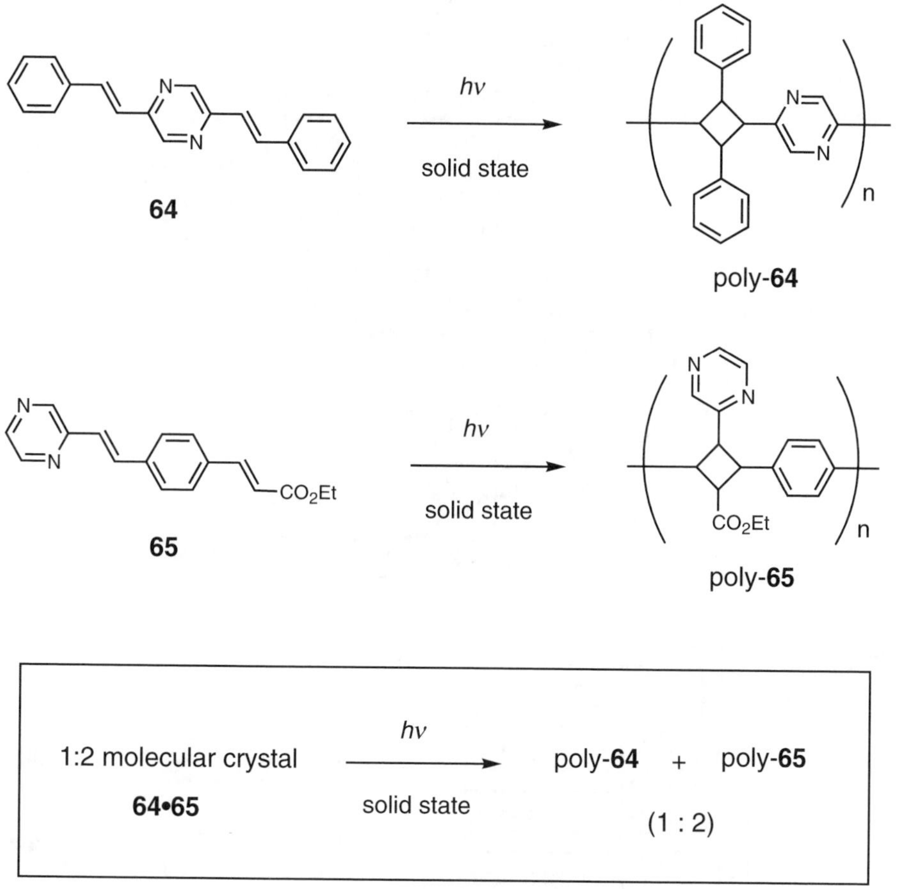

Scheme 6.1 [2 + 2] Photopolymerization in the one-component crystals and two-component crystal of diolefinic compounds.

column of **65**, respectively, i.e. distances are within the limit allowed for topochemical [2 + 2] photocycloaddition [2]. Hence this leads to the production of alternating monolayers of poly-**64** and a bilayer of poly-**65**, as shown in Figure 6.1b.

6.2. [2 + 2] Dimerization of substituted chlorocinnamic acids

One-component crystals of planar chlorocinnamic acids such as 6-chloro-3,4-(methylenedioxy)-cinnamic acid **66**, 2,4-dichlorocinnamic acid **66a** and 3,4-dichlorocinnamic acid **66b** give the corresponding β-truxinic acid dimers as topochemical [2 + 2] photocyclization products **67**, **67a** and **67b**, respectively (**Scheme 6.2**) [46]. A 2:1 two-component crystal **66•66a** was prepared by crystallization from the solution of both

(a)

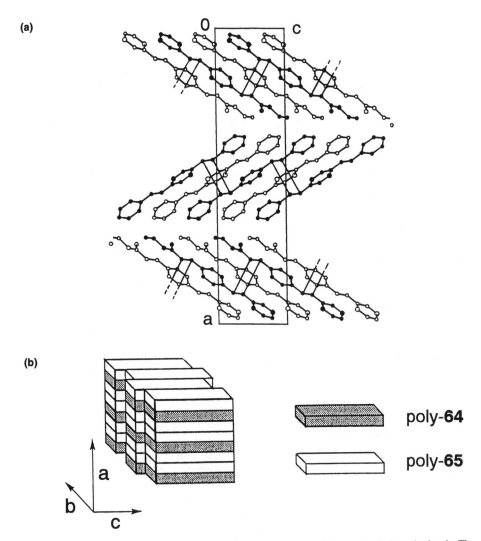

(b)

Figure 6.1 (a) Molecular arrangement in the 2:1 two-component crystal of **64** and **65** along the *b*-axis. The double bonds which form cyclobutanes are linked by solid lines. (b) Schematic representation of the photo-product. Black and white rods represent a single polymer chain of poly-**64** and poly-**65**, respectively. Three arrows represent the direction of the *a*-, *b*- and *c*-axes of the original crystal before irradiation. *Source*: from ref. 45 with permission. © 1993 Am. Chem. Soc.

components in ethanol. A 1:1 crystal **66·66b** (mp 217–218 °C) was also obtained. Solid-state photoreaction of **66·66a** gave mixtures of **67**, **67a** and **67c** in the ratio 4:1:4, respectively. Similarly, **66·66b** gave **67**, **67a** and **67d** in the ratio 1:1:2, respectively. This result is derived from the molecular arrangements that each sheet, which is stabilized by inter-stack C–H•••O, O–H•••O, Cl•••Cl, C–H•••Cl, and O–H•••Cl interactions, is ordered with 2:1 or 1:1 stoichiometry but the stacking along the 4-Å axis is random.

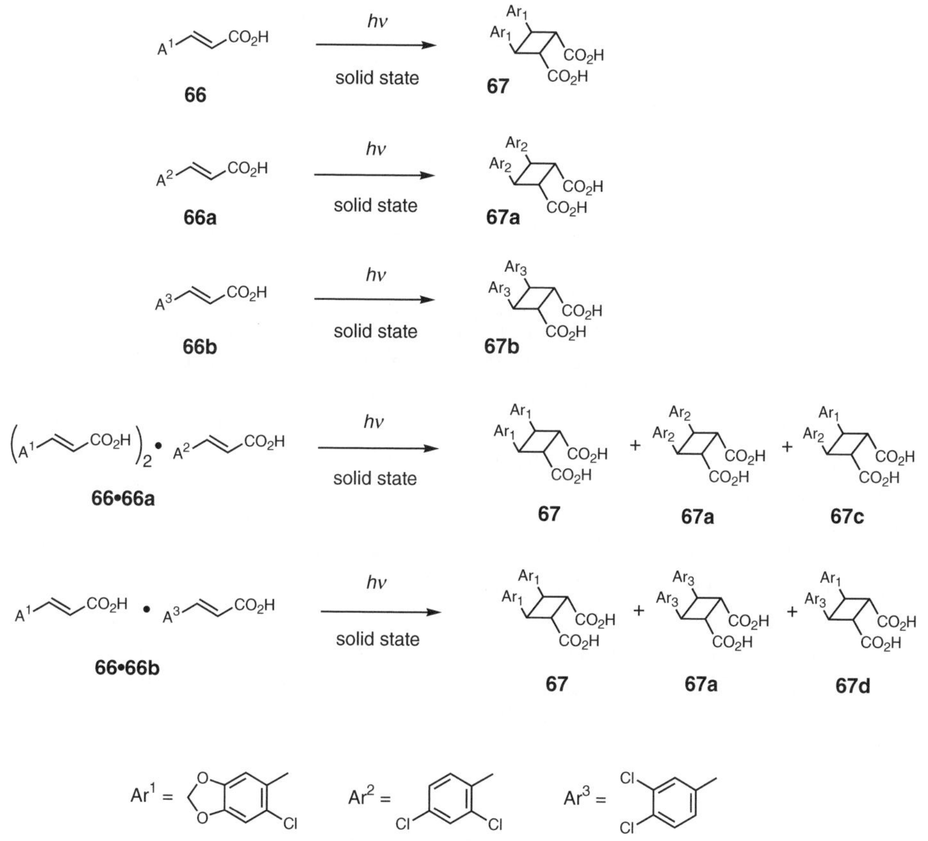

Scheme 6.2 [2 + 2] Photodimerization in the one-component crystals and two-component crystals of cinnamic acids.

6.3. [2 + 2] Cycloaddition in a CT crystal of acenaphthylene and tetracyanoethylene

A tan-brown 1:1 CT crystal **68•69** (mp 103–105 °C) was obtained by crystallization from a 1:1 solution of acenaphthylene **68** and tetracyanoethylene **69** in ethyl acetate. Irradiation of the CT crystal with a high-pressure mercury lamp through a color filter (>500 nm irradiation) at 20 °C caused excitation of the CT band and subsequent [2 + 2] reaction to give a 1:1 cycloadduct **70** in 80% isolated yield as the sole product (**Scheme 6.3**) [47]. The quantum yield for the formation of **70** was about 4 ($\pm$1) $\times 10^{-3}$ under irradiation with 546.1-nm monochromatic light using a thin-layer technique [49]. In contrast, solution reaction of an equimolar (0.1 M) mixture of **68** and **69** in acetonitrile or dichloroethane at >500 nm did not result in any reaction. However, selective excitation of **68** by irradiation at >400 nm led to the formation of **70** in low yield accompanied by larger amounts of the dimer of **68**.

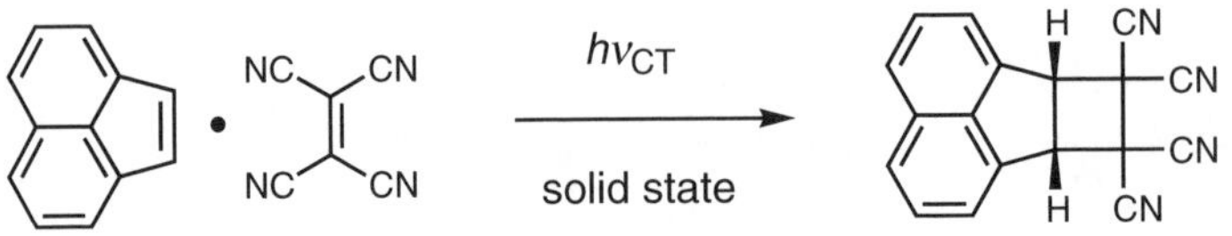

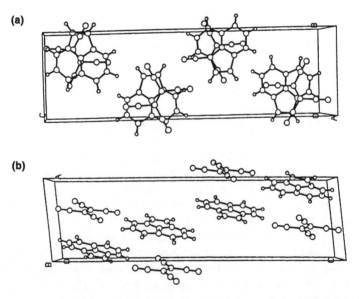

Scheme 6.3 [2 + 2] Cycloaddition in the CT crystal of acenaphthylene and tetracyanoethylene.

The distinctive reactivity and selectivity for product distribution in the crystalline CT complex is attributed to the proximity and rigid mobility of the alkenic C=C double bonds of **68** and **69** in the **68•69** pair. X-ray crystallographic analysis of **68•69** reveals that the molecules of **68** and **69** are stacked alternately and almost parallel with interplanar distance of 4.0 Å to form one-dimensional column structure along the *a*-axis: this is a typical CT arrangement (Figure 6.2a and b). In the molecular pair, two alkenic parts of **68** and **69** are aligned parallel with a torsion angle of 62.2° such that the interatomic distances of C1–C13 and C2–C14 of 3.43 and 3.65 Å, respectively (Figure 6.2c). This molecular arrangement satisfies the requirements for the [2 + 2] cycloaddition between **68** and **69** under topochemical control in the crystal lattice (Schmidt's 4-Å rule) [2]. Another combination of the alkenic carbon atoms, C1–C14 and C2–C13 with distances of 3.97 and 3.33 Å, respectively, is also possible for the cycloaddition, which also satisfies Schmidt's rule. Considering the more remote distance of C1–C14, the former

Figure 6.2 Molecular arrangement in the CT crystal of acenaphthylene and tetracyanoethylene along (a) the *a*-axis and (b) the *b*-axis. *Source*: from ref. 47b with permission. © 1998 Am. Chem. Soc.

(c)

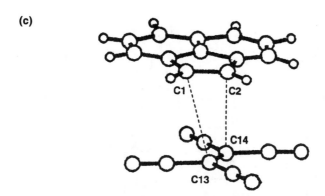

Figure 6.2 (c) the molecular pair.

combination is more favorable than the latter. Topochemical transformation is necessarily accompanied by transfer of the torsion angle of 62.2° between the two alkenic C=C double bonds to the parallel C–C single bond of the cyclobutane moiety in **69**. Although this movement generates large unoccupied space in the crystal lattice, the significantly lower density (1.268 g cm^{-3}) of the crystalline CT complex can satisfy this requirement.

6.4. Photoinduced coupling in the CT crystals of diarylacetylenes and quinone

The crystalline CT complexes incorporate diarylacetylenes (**DA**) such as tolyl-*m*-xylylacetylene (**TX**) as donor and 2,6-dichlorobenzoquinone (**DB**) as an acceptor (**Scheme 6.4**) [49]. The CT crystals were prepared by slow evaporation of a 1:2 solution of diarylacetylene and 2,6-dichlorobenzoquinone in dichloromethane at −20°C in the dark. The CT bands of the crystals were observed at around 500 nm. The molecular arrangements of three representative CT crystals of **DB** with **TX**, **XX** and **FP** are very similar (Figure 6.3). In all three structures, alternating stacks of donor acetylenes and quinone are found, in which there is face-to-face packing of the two quinone molecules with each of the aryl groups on the acetylene. The distances between the planes defined by the aromatic moiety and the quinones are similar: 3.36 and 3.39 Å for **TX**; 3.38 and 3.39 Å for **XX**; and 3.42 and 3.34 Å for **FP**.

The crystals were irradiated with a medium-pressure mercury lamp with a UV cut filter (>410 nm irradiation) under argon at −60°C. Symmetrical acetylene **XX** gave an adduct **71XX** in good yield, while unsymmetrical acetylenes gave the two isomeric mixtures of quinone methides **71a** and **71b** (**Scheme 6.4**). The crystals **FP•DB** with phenyl(pentamethylphenyl)acetylene **FP** had no photoreactivity. Excitation of the CT bands by irradiation at >530 nm afforded similar results to those using >410-nm irradiation.

Transient absorption spectra of the red crystals **FP•DB** dispersed in silica gel were measured by the excitation at 355 nm. Figure 6.4a shows the instantaneous formation of a transient spectrum consisting of an absorption band with $\lambda_{max} = 500$ nm and a

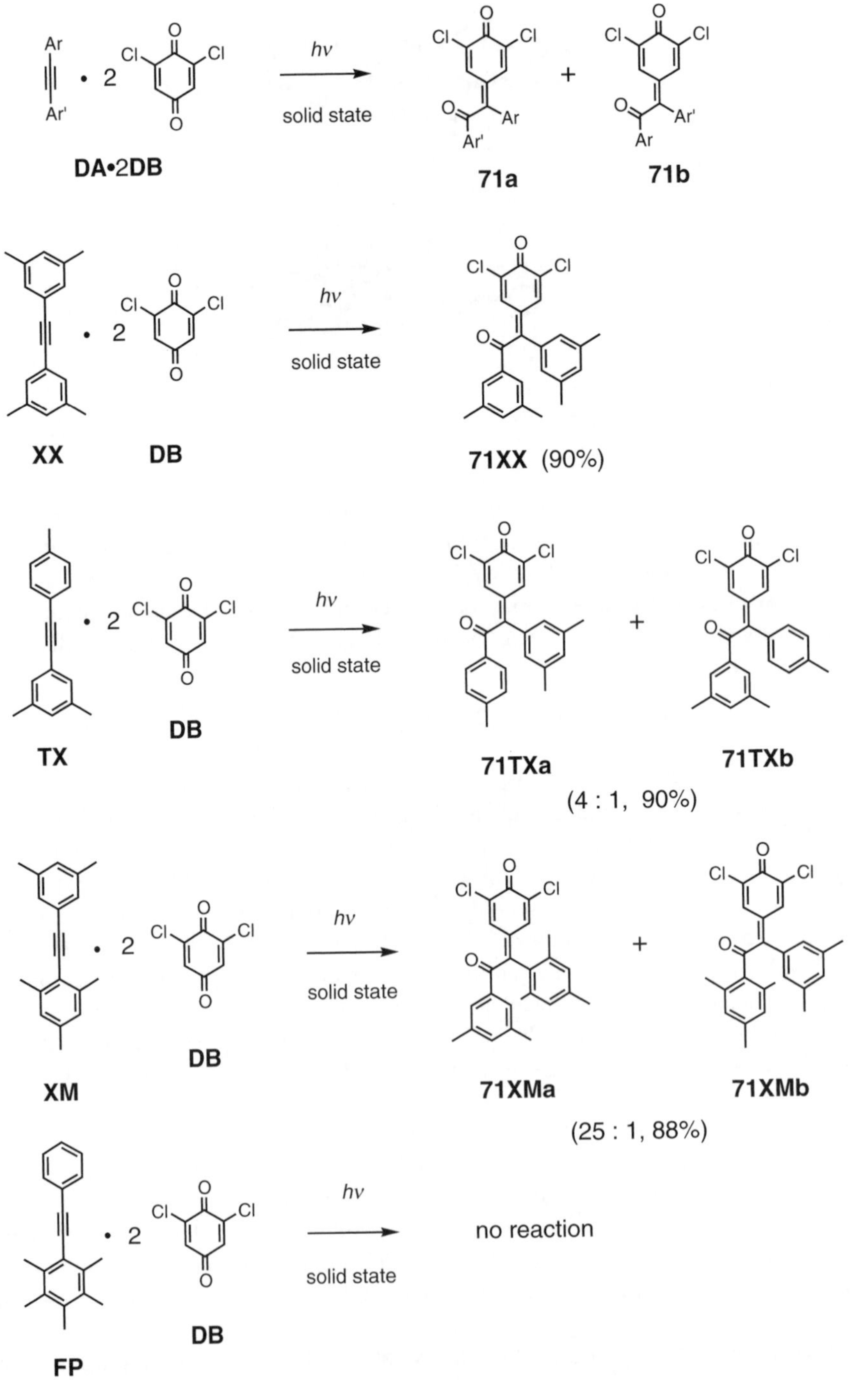

Scheme 6.4 Photoinduced reaction in the CT crystals of diarylacetylenes and dichlorobenzoquinone.

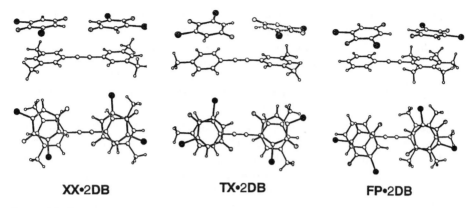

XX•2DB   **TX•2DB**   **FP•2DB**

Figure 6.3 Two perspective views (side and top) of the 1:2 CT crystals of diarylacetylenes and dichlorobenzoquinone. *Source*: from ref. 49 with permission. © 1998 Am. Chem. Soc.

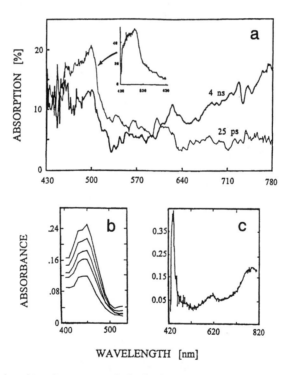

Figure 6.4 (a) Transient absorption spectrum obtained at 25-ps (thin line) and 4-ns (thick line) following the application of the 25-ps laser pulse at 355 nm to the **FP•2DB** crystal. Inset: Transient absorption spectrum of triplet **DB*** obtained from the 25-ps laser pulse at 355 nm of **DB**. For comparison, the authentic spectrum of (b) the anion radical **DB**$^{-•}$ and (c) the cation radical **FP**$^{+•}$. *Source*: from ref. 49 with permission. © 1998 Am. Chem. Soc.

shoulder at 470 nm, which was assigned to the triplet state of quinone by comparison with analogous spectra of other triplet-excited quinones in solution and in the solid state. The lifetime was 4 ns, and then a new absorption band at $\lambda_{max} = 450$ nm and a broad absorption extending beyond 800 nm were observed, which were assigned to the quinone anion radical **DB⁻•** and the acetylene cation radical **FP⁺•** by comparison with authentic samples in Figure 6.4b and c, respectively. On the other hand, the CT excitation at 532 nm gave directly the ion-radical pair.

The reaction mechanism proposed is shown in **Scheme 6.5**. The acetylene cation radical **DA⁺•** and the quinone anion radical **DB⁻•** produced by irradiation are synchronously [2 + 2] coupled and cyclized to give the oxetene intermediate **72**, and then fast rearrangement affords the adduct **71**. However, two questions that remained are for the unreactivity of the CT crystal with **FP** and the regioselectivity in the coupling products **a** and **b**, despite of similar molecular arrangements among the three crystals (Figure 6.3). The distances between the respective reactive centers (i.e., the triplet bond of the donor and the carbonyl groups of the two quinones) are not very different, 3.65 Å (dimethyl aromatic ring) and 4.15 Å (methylphenyl ring) for **TX**, 3.6 Å for **XX**, and 4.34 Å (pentamethylphenyl ring) and 3.84 Å (phenyl ring) for **FP**. The angles between the two quinone O–O axes and the long axis of the acetylene donor are also similar; 36.8° and 37.6° for **TX**, 35° for **XX**, and 52.9° and 41.3° for **FP**. At this time, the detailed analysis based on the least-motion postulate requires further evaluation.

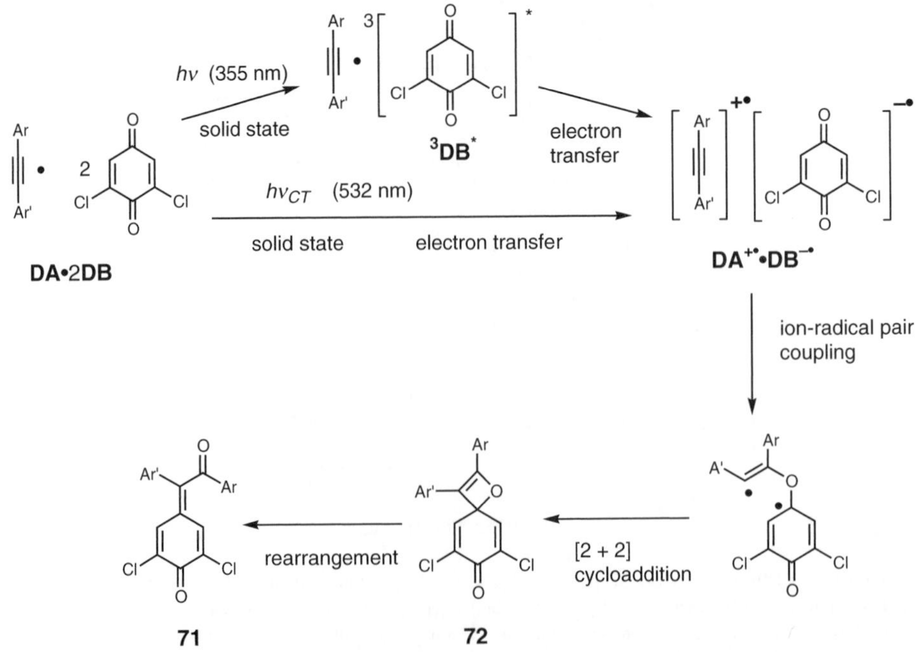

Scheme 6.5 Reaction mechanism in the CT crystals of diarylacetylenes and dichlorobenzoquinone.

7. Substitution

The guest ethanol molecules in the clathrate crystals reacted photochemically with the diol host compounds, bis(9-hydroxyfluoren-9-yl)thieno[3,2-b]- and [2,3-b]-thio-phenones **73** and **74**, to cause photosolvolysis in the solid state [50,51].

Recrystallization of **73** from ethanol solution gave 1:2 host-guest compound, while a host-guest ratio of 2:1 for **74** was obtained. Irradiation of the pulverized crystals of **73**•(EtOH)$_2$ with a high-pressure mercury lamp at ambient temperature caused photo-substitution with the guest ethanol molecule to afford monoether **75a** (43%) and diether **75b** (21%) along with unreacted **73** (36%) (**Scheme 7.1**). Similarly, photosubstitution of (**74**)$_2$•EtOH gave monoether **76a** (36%) and diether **76b** (7%). It was confirmed that the crystalline phase is retained throughout the photoreaction by the measurement of the X-ray powder diffraction patterns of both irradiated crystals.

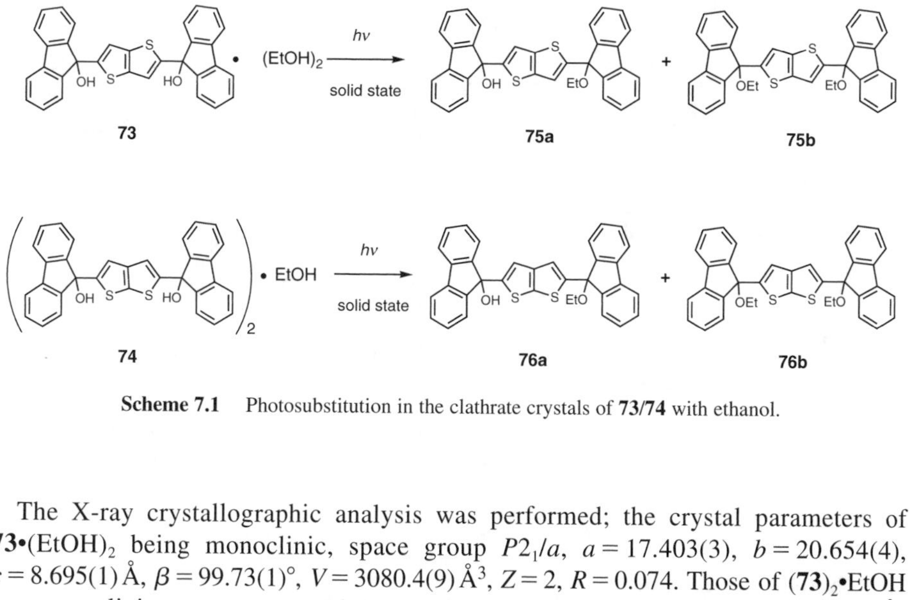

Scheme 7.1   Photosubstitution in the clathrate crystals of **73/74** with ethanol.

The X-ray crystallographic analysis was performed; the crystal parameters of **73**•(EtOH)$_2$ being monoclinic, space group $P2_1/a$, $a = 17.403(3)$, $b = 20.654(4)$, $c = 8.695(1)$ Å, $\beta = 99.73(1)°$, $V = 3080.4(9)$ Å^3, $Z = 2$, $R = 0.074$. Those of (**73**)$_2$•EtOH are monoclinic, space group $P2_1/c$, $a = 24.359(3)$, $b = 13.793(2)$, $c = 15.508(1)$ Å, $\beta = 99.728(8)°$, $V = 5135(1)$ Å^3, $Z = 2$, $R = 0.094$. The packing arrangements revealed that the ethanol molecules are connected through the hydrogen bonding between each of the OH groups of the host molecules (Figure 7.1). The interatomic distances between the ethanol oxygen atom and the fluorenyl 9-carbon atom indicate that 3.65 and 3.86 Å in **73**•(EtOH)$_2$ as well as 3.73 and 3.74 Å in (**74**)$_2$•EtOH are possible distances for dis-placement to give **75a**, **75b** and **76a**. However, for the formation of **76b** from (**74**)$_2$•EtOH, the local microenvironmental concentration of ethanol surrounding the OH group of host **74** is insufficient, since the bond formation across a distance as long as 5.06 Å is required (Figure 7.1). This distance appears to be too long for efficient topochemical solid-state reactions, leading to the low yield (7%) of **76b**.

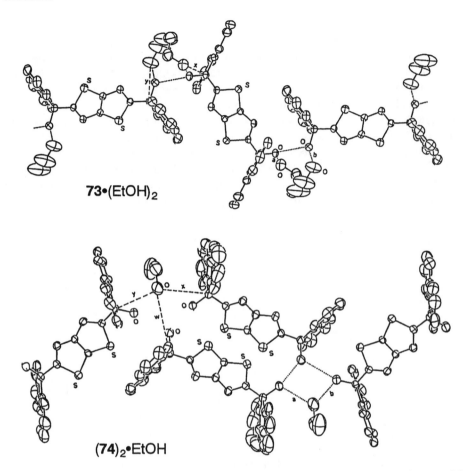

73•(EtOH)₂

(74)₂•EtOH

Figure 7.1 ORTEP views of the clathrate crystals **73•(EtOH)₂** and **(74)₂•EtOH**. Dotted lines a and b represent the relation linked by the hydrogen bond between the host and the guest, and broken lines x, y and w represent the direction of the new bond formations. Hydrogen atoms are omitted for clarity. *Source*: from ref. 50 with permission. © 1994 Elsevier Science.

8. Photodecarboxylation

Photodecarboxylations of organic carboxylic acids are well known to occur via photo-induced electron transfer [52,53]. Although a large number of photodecarboxylations in the solution phase have been studied by using various acceptors such as aza aromatic compounds [54] and polycyanoaromatics [55], the product selectivities are necessarily low due to the co-occurrence of subsequent radical coupling because of the free motion of radical species in solution. In contrast, solid-state photoreaction in two-component crystals combining various carboxylic acids and electron acceptors occur with a remarkable selectivity and specificity compared with the solution-phase photodecarboxylations [57–62]. Stoichiometrically sensitized decarboxylation is observed in a series of two-component crystals of aralkyl carboxylic acids and aza aromatic compounds to give

230

decarboxylated product alone quantitatively [56]; photodecarboxylation and decarboxylative condensation of arylacetic acids by excitation of the CT bands of the complexes with 1,2,4,5-tetracyanobenzene in the crystalline state [57]; retention of the initial chirality of (S)-(+)-2-(6-methoxy-2-naphthyl)propanoic acid during decarboxylative condensation in its CT crystal with TCNB [58]; decarboxylation and dehydrogenation of α-hydroxycarboxylic acids in their two-component crystals with aza aromatic compounds [59]. Ultimately, even absolute asymmetric synthesis was achieved via photodecarboxylative condensation in a chiral two-component crystal formed from achiral molecules of acridine and diphenylacetic acid (see Section 9.3) [63].

8.1. *Stoichiometrically sensitized decarboxylation*

Remarkably selective photodecarboxylation occurs in a series of two-component crystals combining aralkyl carboxylic acids with aza aromatic compounds as electron acceptors. On the basis of the correlation between the crystal structures and the reactions, a new concept of stoichiometric sensitization was established, which is specific in the solid state photoreaction [56].

The series of two-component crystals incorporate aralkyl carboxylic acids such as 3-indoleacetic acid **a**, 1-naphthylacetic acid **b**, 9-fluorenecarboxylic acid **c**, and 9-fluoreneacetic acid **d**, and aza aromatic compounds such as acridine **52** and phenanthridine **77** as electron acceptors (**Scheme 8.1**). Deuterated 1-naphthylacetic acid **bD**, of which the CO_2H group was exchanged to CO_2D, was also used. Crystallization from solutions of both components in appropriate solvents gave the two-component crystals. The molar ratio of all the crystals was 1:1.

X-ray crystallographic analysis revealed that O–H•••N hydrogen bonds (2.6–2.7 Å) were formed between the carboxyl group and the N atom of **52** or **77** in all the crystals. In the case of the crystals **52•a** and **77•a** with 3-indoleacetic acid, additional N–H•••O–C hydrogen bondings (2.0–2.1 Å) are formed between the indole imino group and the carboxyl group of the adjacent indole ring. Figures 8.1–8.3 show the molecular arrangements of representative crystals. In the unit cell of **77•a** from phenanthridine and 3-indoleacetic acid, two hydrogen-bonded pairs of **77** and **a** are stacked in head-to-tail manner to form a linear chain structure among the molecules through the N–H•••O–C hydrogen bonding (Figure 8.1). In the crystal **52•b** from acridine and 1-naphthylacetic acid, hydrogen-bonded pairs of **52** and **b** are stacked head-to-tail and four pairs are packed per unit cell (Figure 8.2). Surprisingly, the crystal **52•bD** of which CO_2H was replaced by CO_2D changed the crystal system from monoclinic (**52•b**) to triclinic (**52•bD**). Nevertheless, the molecular packing arrangements in head-to-tail pairs are rather similar. In the case of **72•c**, the hydrogen-bonded pairs arrange in head-tead manner to form a column structure in the crystal lattice (Figure 8.3).

Solid-state irradiation of the pulverized two-component crystals with a xenon short arc lamp through a UV transparent filter (290–400 nm irradiation) under argon at -70°C caused highly selective decarboxylation to give the corresponding decarboxylated product **78** in high yield (**Scheme 8.2** and Table 8.1). The decarboxylative condensation product **79** or the dehydrogenated product **80** were obtained as the minor products. In

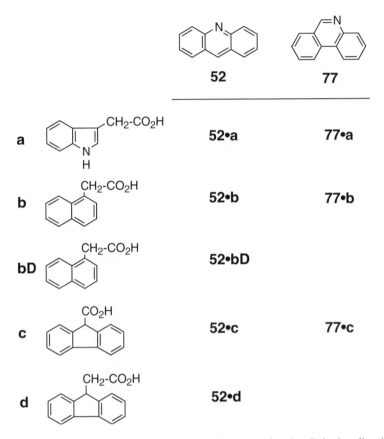

Scheme 8.1   Two-component crystals of aza aromatic compounds and aralkylcarboxylic acids.

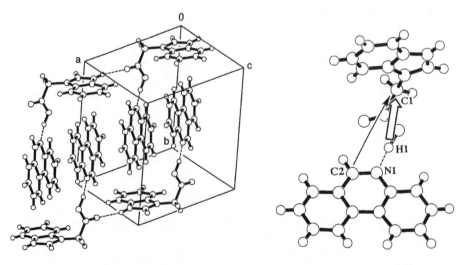

Figure 8.1   Molecular arrangement and the radical coupling path in the two-component crystal **77•a**.

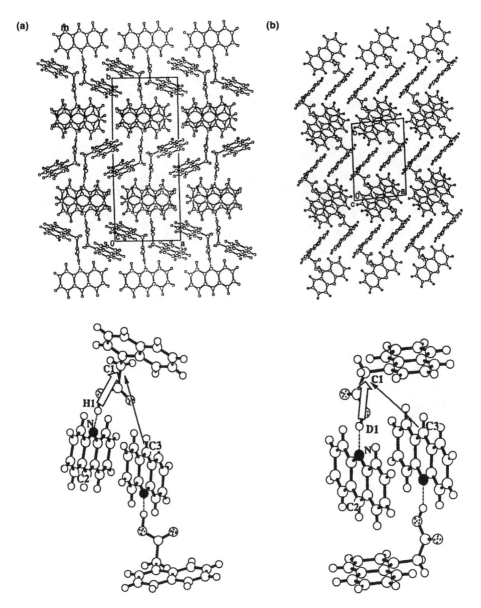

Figure 8.2   Molecular arrangements and their radical coupling paths in the two-component crystals (a) **52•b** and (b) **52•bD**. *Source*: from ref. 56 with permission. © 1997 Am. Chem. Soc.

particular, **52•c**, **52•d**, **77•a** and **77•c** underwent completely selective decarboxylation to give the corresponding decarboxylated product **78** alone. Irradiation at 15 °C decreased the selectivity, e.g., **77•a** gave **78a** (69%) and **80a** (13%). Thus lowering the irradiation temperature led to an increase in the selectivities due to a smaller thermal motion of the

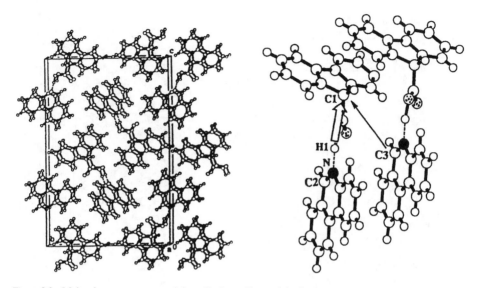

Figure 8.3 Molecular arrangement and the radical coupling path in the two-component crystal **77•c**. *Source*: from ref. 56 with permission. © 1997 Am. Chem. Soc.

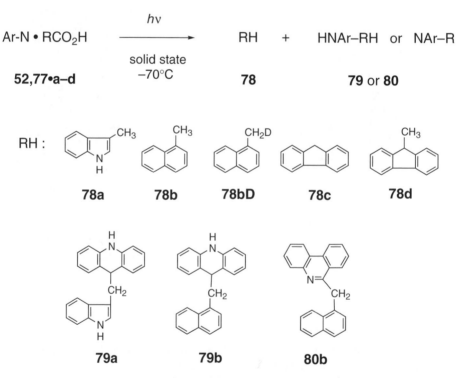

Scheme 8.2 Solid-state photodecarboxylation in the two-component crystals of aza aromatic compounds and aralkyl carboxylic acids at −70°C.

Table 8.1 Solid-state photoreaction in the two-component crystals of aza aromatic compounds and aralkyl carboxylic acids at $-70\,^{\circ}$C

Entry	Crystal	Conversion (%)		Yield (%)	
		52 or **77**	acid	decarboxylated product **78**	condensation product
1	**52•a**	12	10	90	5 (**79a**)
2	**52•b**	19	50	81	4 (**79b**)
3	**52•bD**	9	30	88[a]	3 (**79b**)
4	**52•c**	4	5	94	0
5	**52•d**	4	12	93	0
6	**77•a**	0	23	92	0
7	**77•b**	5	35	93	4 (**80b**)
8	**77•c**	0	30	98	0

[a] 1-Methyl(CH$_2$D)naphthalene.

radical species. Furthermore, the acceptor aza aromatic compounds **52** or **77** were completely recovered without any consumption.

In contrast, solution-phase photolysis of aza aromatic compounds and carboxylic acids in acetonitrile with a high-pressure mercury lamp under argon at room temperature caused similar decarboxylation followed by radical coupling to give typically mixture of four products (**Scheme 8.3** and Table 8.2). These are the decarboxylated product **78**, decarboxylative condensation product **79**, the dimer **82** of the decarboxylated product, and the dimeric biacridane **83** or biphenanthridane **84** of aza aromatic compound. In some cases, the dehydrogenated condensation product **80** or **81** was obtained. The selectivities were low, which is very different from those in solid state. Specifically, all the possible radical coupling products formed in low selectivities due to the high mobility of intermediate radical species in the solution.

The possible reaction mechanism based on photoinduced electron transfer is shown in **Scheme 8.4**. Irradiation of a crystal excites **52** or **77**, followed by electron transfer from the acid to **52** or **77**, to afford an anion radical and a cation radical. Then proton transfer can give a carboxylate radical **85**, a hydroacridine radical **86** or hydrophenanthridine radical **87**, which are represented by their most favorable canonical forms. Next, decarboxylation of **85** produces the •CH$_2$– aralkyl radical **88**. Finally, hydrogen abstraction can occur from the N–H of **86** or **87** by the reactive radical **88**, to result in formation of the decarboxylated product **78** and the regeneration of **52** or **77**, while radical coupling between **15** and **86** or **87** gives the condensation product **79**.

Although these processes inevitably lead to alteration of the crystal lattice, the radical species can move a little during their lifetimes leading to hydrogen abstraction and radical coupling. The reaction paths can be rationalized using crystallographic data pertaining to the two-component molecular crystals such as the estimated distances in Table 8.3. The occurrence of hydrogen abstraction as a higher priority than radical coupling can be explained from the shorter distances available for hydrogen abstraction (3.2–3.5 Å) than those for radical coupling (4.5–6.5 Å). Furthermore, the hydrogen atom is very small in

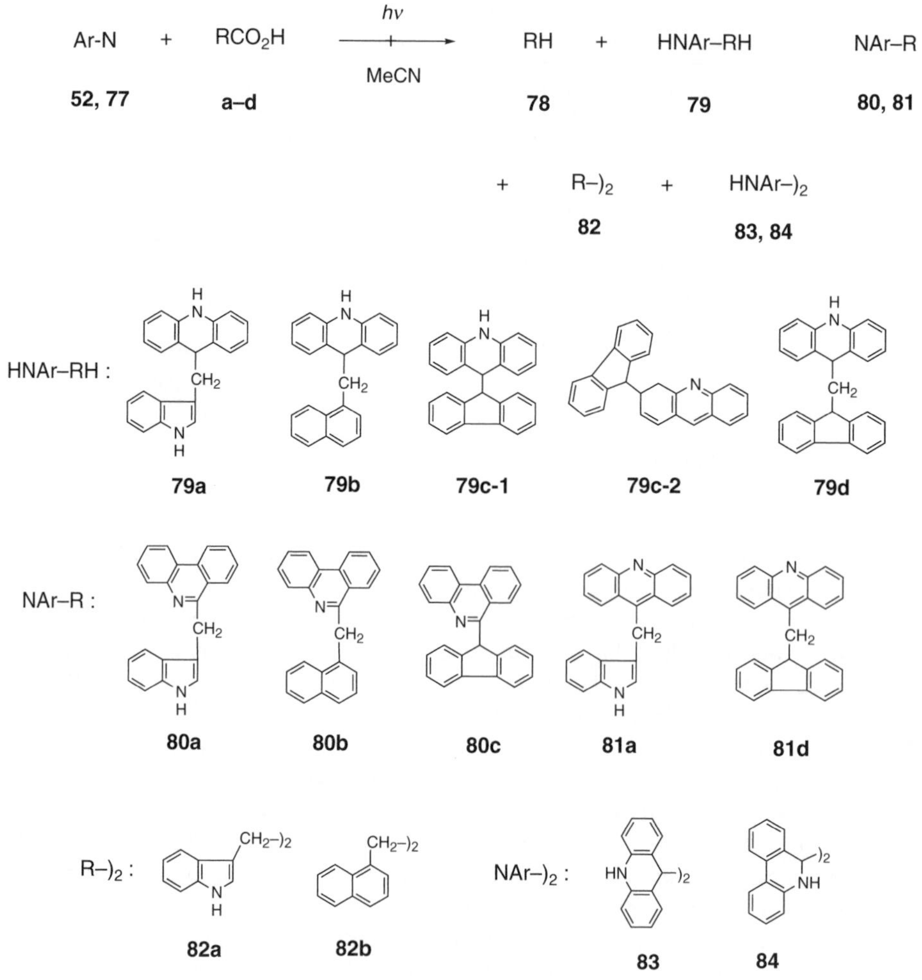

$$\text{Ar-N} + \text{RCO}_2\text{H} \xrightarrow[\text{MeCN}]{h\nu} \text{RH} + \text{HNAr–RH} \quad\quad \text{NAr–R}$$

52, 77   **a–d**   **78**   **79**   **80, 81**

$$+ \quad \text{R–})_2 \quad + \quad \text{HNAr–})_2$$

82   **83, 84**

HNAr–RH :

79a   **79b**   **79c-1**   **79c-2**   **79d**

NAr–R :

80a   **80b**   **80c**   **81a**   **81d**

R–)₂ : CH₂–)₂   CH₂–)₂   NAr–)₂ :

82a   **82b**   **83**   **84**

Scheme 8.3 Photolysis of aza aromatic compounds and aralkyl carboxylic acids in acetonitrile.

comparison with the large size of **86** or **87**, implying its higher mobility in the solid state. The radical coupling paths are illustrated in Figures 8.1 (**77•a**), 8.2 (**52•b, 52•bD**) and 8.3 (**77•c**). For instance, in the crystal **77•a**, hydrogen abstraction occurs between radical species **87** (H1) and **88** (C1), which are derived from a hydrogen-bonded pair since the C1—H1 distance (3.18 Å) is the shortest in the crystal lattice (open arrow in Figure 8.1), assuming that the relative positions of these radical species formed in the crystal lattice have not changed much from those of the starting crystals.

On the other hand, radical coupling occurs between a geminate radical pair **88** and **86** or **87**, which are derived from a hydrogen-bonding pair (C1—C2) or neighboring hydrogen-bonded pairs (C1—C3), depending on the molecular arrangement in the crystal lattice. For instance, in the crystal **77•a** the radical coupling can occur between C1 and C2

Table 8.2 Photolysis of aza aromatic compounds and aralkyl carboxylic acids in acetonitrile

Component	Conversion (%)		Yield (%)				
	52 or 77	acid	78	79	80 or 81	82	83 or 84[a]
52+a	80	75	0	49	4	0	45
52+b	81	87	2	70	10	0	52
52+c	86	98	28	43[b], 10[c]	0	0	6
52+d	75	57	21	30	26	0	37
77+a	15	25	14	0	35	35	23
77+b	40	87	17	0	37	9	12
77+c	11	99	85	0	10	0	0

[a] Yield based on consumed aza aromatic compound.
[b] Compound **79c-1**.
[c] Compound **79c-2**.

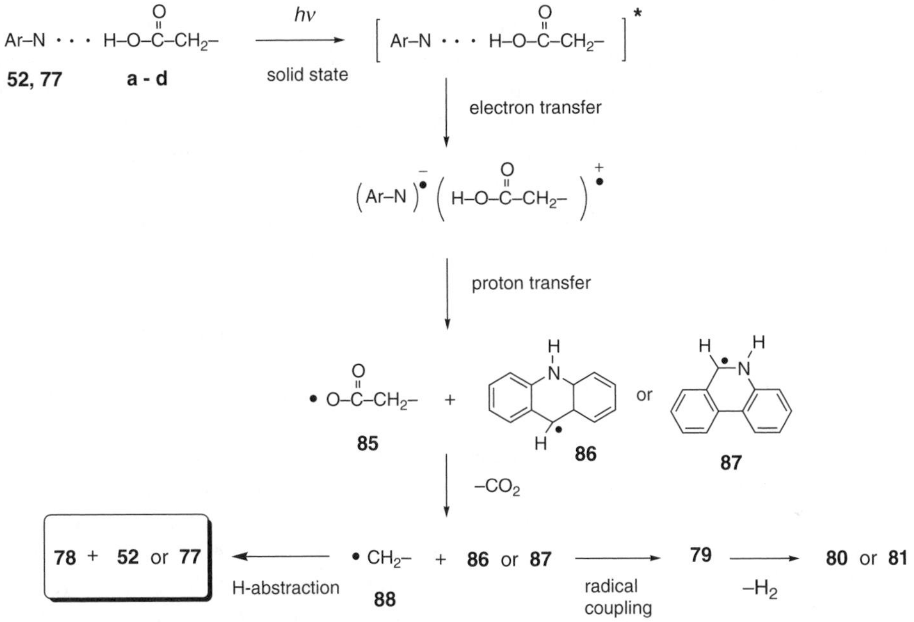

Scheme 8.4   Possible photochemical reaction mechanism in the two-component crystals of aza aromatic compounds and aralkyl carboxylic acids.

within the geminate radical pair derived from a hydrogen bonding pair because the C1—C2 distance (5.34 Å) is shorter than the C1—C3 (8.04 Å) distance due to the head-to-tail stacking (Figure 8.1 and Table 8.4).

Evidence was obtained that the radical **88** eventually abstracted the H1 hydrogen atom of the CO_2H group by irradiating the deuterated crystal **52•bD**. As shown in Table 8.1, **52•bD** similarly caused highly selective decarboxylation to give 1-CH_2D-naphthalene

Table 8.3 Estimated distances for the hydrogen abstraction and radical coupling

Distance[a] (Å)	52•a	52•b	52•bD	52•c	52•d	77•a	77•c
Hydrogen abstraction within H-bonding pair							
C1—H1	3.33	3.47	3.20[b]	3.31	3.30	3.18	3.40
Radical coupling within H-bonding pair							
C1—C2	7.40	7.57	7.75	7.71	7.53	5.34	5.23
Radical coupling between neighboring H-bonding pairs							
C1—C3	5.95	6.06	5.08	6.49	5.41	8.04	4.75

[a] See Figures 8.1–8.3 for the atom numbering.
[b] H is exchanged to D.

(88%), which was confirmed by gas chromatography-mass spectrometry (GLC-MS) and ^{1}H NMR measurements.

Finally stoichiometric sensitization, which is a new concept, was proposed. The acceptor molecules of acridine **52** or phenanthridine **77** are regenerated by hydrogen abstraction from the radical species **86** or **87** in the solid state. Dimerization of **86** or **87** is not caused by their low mobility in the crystal lattice, despite the formation of dimers **83** or **84** in the solution photoreaction. This means that the aza aromatic compounds behave like a sensitizer, acting only in one cycle while retaining the initial crystal structure, i.e. a stoichiometric sensitizer. Such a stoichiometric sensitization is a novel photochemical process, which occurs specifically in the crystalline state.

8.2. *Photodecarboxylation in the CT crystals with 1,2,4,5-tetracyanobenzene*

Crystalline CT complexes of TCNB **26** (as an electron acceptor) and aralkyl carboxylic acids (such as 1-naphthylacetic acid **b**) also cause photodecarboxylation and then dehydrocyanating condensation by CT band excitation (**Scheme 8.5**) [57]. The 1:1 CT crystals **26•b** are obtained from the 1:1 solution of **26** and **b** in acetonitrile. The CT band in the solid state is observed over 370–460 nm. Figure 8.4 illustrates the molecular arrangement in this lattice. The parallel and alternate stacking (3.4 Å) of the molecules **26** and **a**, with superimposition in some orientations, implies that the π–π interaction has CT character. Further, the carboxy dimer of **b** between the neighboring columns is formed through O–H•••O hydrogen bonding (1.83 Å).

Excitation of the CT band by irradiating at >390 nm caused decarboxylation and subsequent dehydrocyanative condensation to give methylnaphthalene **78b** and 1-naphthyl(2,4,5-tricyanophenyl)methane **89b** (**Scheme 8.5**). UV irradiation with a high-pressure mercury lamp through Pyrex glass also gave the two products with similar selectivity and yield.

In comparison, solution-phase photolysis of **26** and **b** in acetonitrile at >290 nm under argon caused decarboxylation to give the condensation product **89b** as the major product and 1-naphthylaldehyde **90b** as the minor product (**Scheme 8.5**). It should be emphasized that no formation of the decarboxylated product **78b** was observed in the

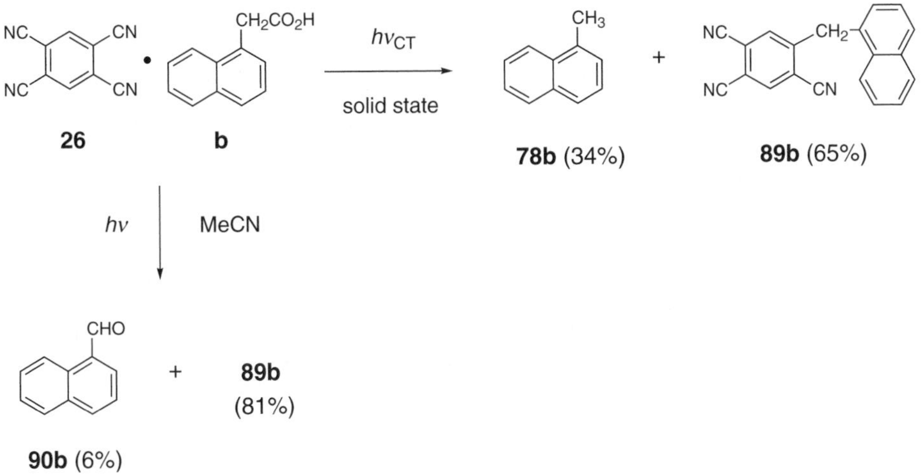

Scheme 8.5 Photoreaction of TCNB and 1-naphthylacetic acid in the CT crystalline state and in solution.

solution photoreaction. The formation of aldehyde **90b** is ascribed to result from partici-
pation of the cation radical species of water present in the solvent acetonitrile. Further,
excitation of the CT band at >390 nm in the solution phase hardly caused any condensation, giving **89b** in only 2% yield. Instead, the aldehyde **90b** (37%) was the major product. The reason for this ineffectiveness is most probably the low concentration of the CT
complex **26•b** (in equilibrium with its separate components) in the solution.

Similar reactions occurred in the CT crystal of TCNB and 2-naphthylacetic acid and
in the solution phase. On the contrary, the CT crystal of TCNB and 3-indoleacetic acid
gave no reaction in the solid state, despite the occurrence of a similar decarboxylating
condensation in acetonitrile.

The possible reaction pathway of the solid-state photoreaction initiated by excitation
of the CT crystal has been proposed (**Scheme 8.6**). Excitation of the CT band initially
causes photoinduced electron transfer to give the carboxylic acid cation radical and
TCNB anion radical. The condensation product **89b** is produced by the formation of
coupled intermediate **92** from a radical •CH$_2$Ar and an anion radical **91**, followed by the
elimination of a cyanide anion. The formation of **78b** is attributed to the hydrogen transfer between a radical •CH$_2$Ar and radical intermediate **93**.

It is understandable from Figure 8.4 that CT excitation causes electron transfer from
the acid to TCNB **26** along the stacking column to give the radical cation of acid and the
radical anion **91**. The shortest •C—C distance between the decarboxylated radical
•CH$_2$Ar and the phenyl ring of **26** is estimated as 3.75 Å (C1—C2) from the crystal data.
This distance is short enough to allow radical coupling giving the intermediate **92** and
subsequently producing **89b**. Since the other •C—C distances are 4.61 Å (C1—C3),
4.62 Å (C1—C4), and 4.68 Å (C1—C5), the possibilities of the radical coupling are
lower. On the other hand, the production of **78b** can be explained as follows. Despite the
slightly longer H1—C2 distance of 4.31 Å, the active radical cation **91** can abstract H$^+$
to give the intermediate **93**, from which the radical •CH$_2$Ar subsequently abstracts a

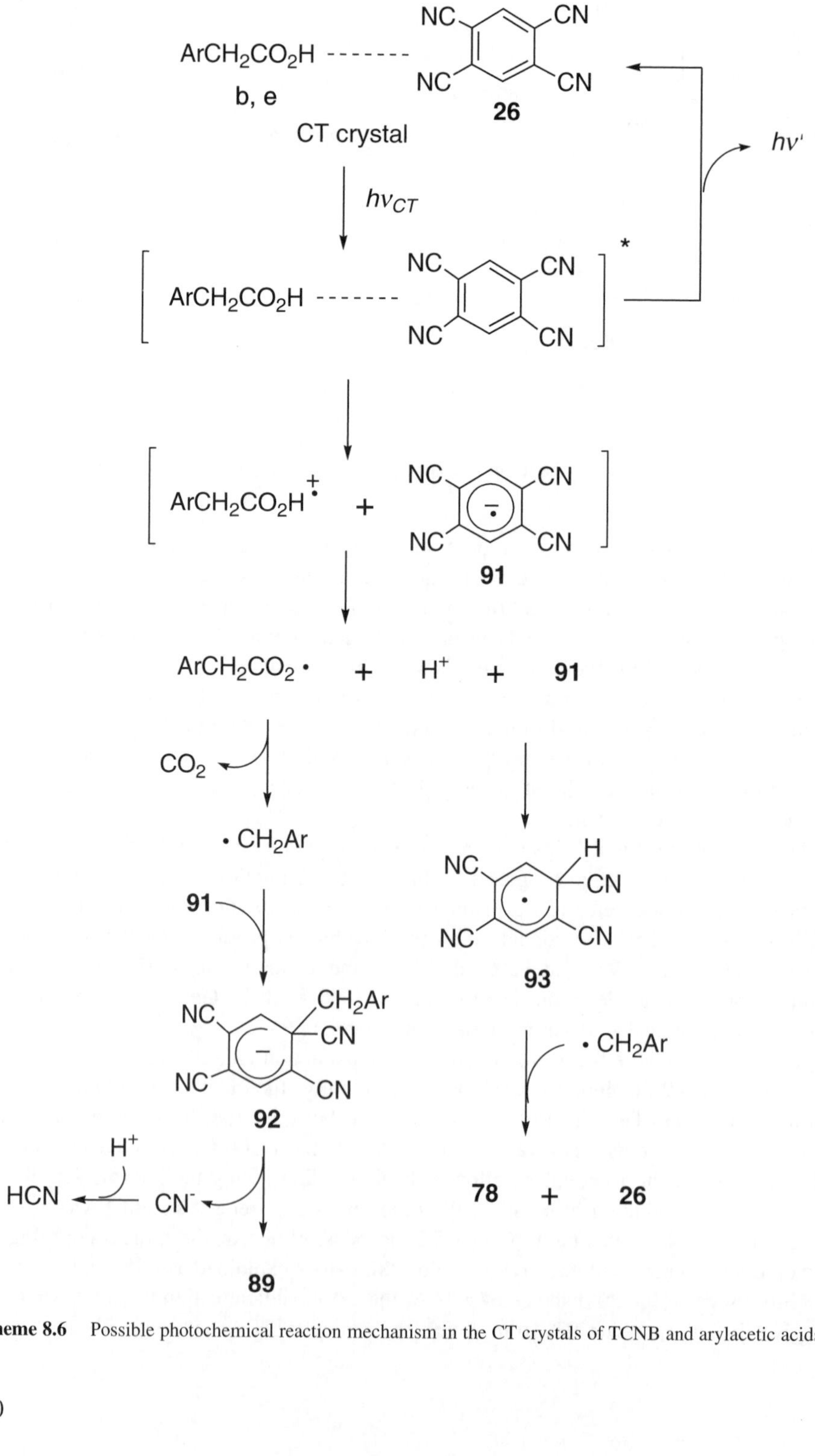

Scheme 8.6 Possible photochemical reaction mechanism in the CT crystals of TCNB and arylacetic acids.

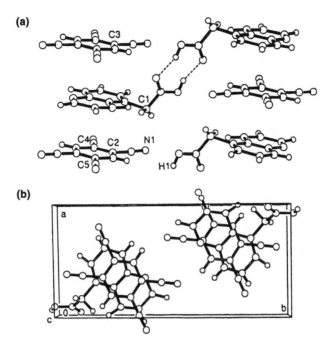

(a)

C3

C1

C4 C2 N1
C5 H1

(b)

a

c b

Figure 8.4 Molecular arrangement in the CT crystal of TCNB and 1-naphthylacetic acid. *Source*: from ref. 57 with permission. © 1996 Am. Chem. Soc.

hydrogen atom to produce **78b**. Accordingly, these reactions producing **78b** and **89b** are competitive.

The quantum yields (Φ) for the photoreaction of the CT crystals **26•b** at 300–330 nm irradiation were measured using the thin-layer technique [48] to give Φ values for **78b** (0.043) and **89b** (0.10). The Φ values for **89b** (0.05) and **89e** (0.23) in the acetonitrile solution of the two components using the usual merry-go-round technique were comparable to that in the solid state. The relatively high Φ value for the aldehyde **90b** (0.17) formation in acetonitrile compared to the final preparative yield of 6% (**90b**) is attributable to its facile production in the early stages of the photoreaction.

8.3. Chirality memory

Crystalline-state reaction of CT complex of TCNB **26** and (*S*)-(+)-2-(6-methoxy-2-naphthyl)propanoic acid (*S*)-(+)-**94** caused enantioselective radical coupling with retention of the initial configuration (**Scheme 8.7**) [58]. The CT crystal was prepared from an equimolar mixture of **26** and (*S*)-(+)-**94** in 3:1 acetonitrile-methanol (mp 168 °C), for which a CT band was observed at 433 nm as a broad band. Irradiation of the chiral CT crystals caused decarboxylation and then enantioselective radical coupling to give the optically active decarboxylative condensation product (+)-**95** (**Scheme 8.7**). The optical rotation $[\alpha]_D$ [20] was +22 (*c* = 0.30, MeOH) and the optical purity is 21% ee for 95%

241

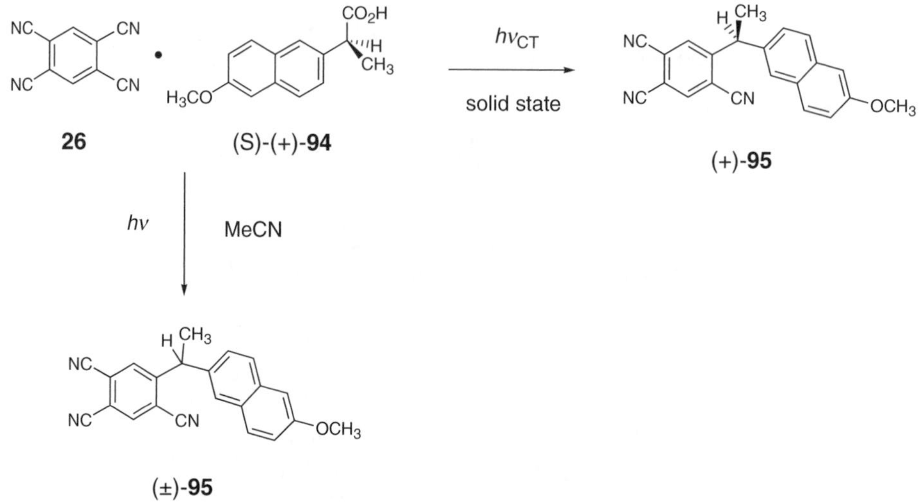

Scheme 8.7   Photoreaction of TCNB and (S)-(+)-(6-methoxy-2-naphthyl)propanoic acid in the CT crystalline state and in solution.

chemical yield, which imply that the chirality of **94** is transferred to product **95** in the decarboxylative coupling process. The reaction probably proceeds via the same radical coupling path shown in **Scheme 8.6**. The radical species (Ar–MeHC•) formed in the solid-state reaction may retain its original hybridization state and configuration to some extent because of the restrictions of the crystal lattice when it reacts with the TCNB anion radical **91** to form a new C–C bond.

In contrast, while solution photolysis between **26** and (S)-(+)-**94** in acetonitrile gave the racemic condensation product (±)-**95** in 72% yield, both the optical rotation and enantiomeric excess were found to be zero (**Scheme 8.7**). This suggests that the radical species (Ar–MeHC•) undergoes change of hybridization from sp^3 to sp^2 and loses its original configuration in the free environment of the solution phase before it reacts with the TCNB anion radical **91** to form a new C–C bond.

Figure 8.5 illustrates the enantiomeric molecular arrangement in the CT crystal, which belongs to chiral space group *P*1. The alternate layer stacking of **TCNB** and **94** shows the CT character with a plane-to-plane distance of 3.4 Å. The shortest coupling distances between the radical Ar–MeHC• and the TCNB anion radical **91** are estimated to be 4.68 (C1—C2), 4.81 (C1—C3), 4.83 (C1—C4) and 5.24 Å (C1—C5) based on the crystal data. Despite these being rather long for radical coupling, the formation of optically active (S)-(+)-**94** suggests that the radical species can move in the crystal lattice.

8.4. Enantioselective photodecarboxylative condensation

Enantiomorphous two-component crystals from acridine and chiral aralkyl carboxylic acids give enantiomeric products as well as diastereomeric products through enantioselective

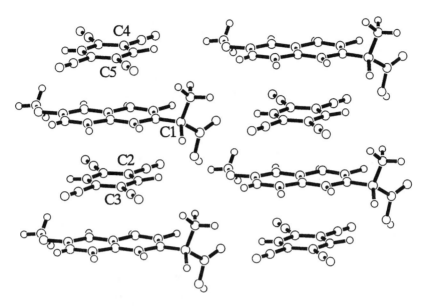

Figure 8.5   Molecular arrangement in the chiral CT crystal formed between (*S*)-(+)-**94** and TCNB.

photodecarboxylative condensation (**Scheme 8.8**) [59]. The enantiomorphous crystal **52•R** was prepared by crystallization of a 3:2 ratio of acridine **52** and (*R*)-(−)-2-phenyl-propionic acid **R** in ethanol. The melting point is 73 °C, appearing between those of **52** (179 °C) and **R** (29 °C). Crystalline **52•R** was also obtained by grinding a 3:2 mixture of **52** and **R** without solvent in a mortar for a several tens of minutes. The opposite-handed crystal **52•S** was similarly prepared by crystallization from solution or grinding of **52** and (*S*)-(+)-2-phenylpropionic acid **S**.

Figure 8.6a illustrates the structure of 3:2 crystal **52•R**, which crystallizes in space group *P*2₁. The absolute configuration was determined using the reference of **R**. Two independent hydrogen-bonded pairs of **52** and **R** are formed with the O–H•••N distances of 1.57 Å and 1.89 Å and with a dihedral angle between the acridine and carboxyl planes of 34.8° and 13.8°, respectively. The remaining aciridine molecule does not hydrogen bond. In total, four hydrogen-bonded pairs plus two acridine molecules are present per unit cell, and further stacking along the *a*-axis forms a columnar structure.

Irradiation of crystalline **52•R** caused decarboxylation and enantioselective condensation giving three optically active products **96**, **97**, and **98** with positive [α]_D [20] values (**Scheme 8.8**). Similarly, the opposite-handed crystal **52•S** afforded **96**, **97**, and **98** with negative [α]_D [20] values. In contrast, photolysis of a 1:1 solution of **52** and **R** resulted in formation of optically inactive **96** and biacridane **83**.

Upon irradiation of **52•R**, electron transfer, proton transfer and then decarboxylation occur to give •CHMePh radical and hydroacridine radical species. Next C1—C2 radical coupling results as the highest priority event due to the short distance of 4.8 Å and the high charge density on the hydroacridine radical. The original configuration of

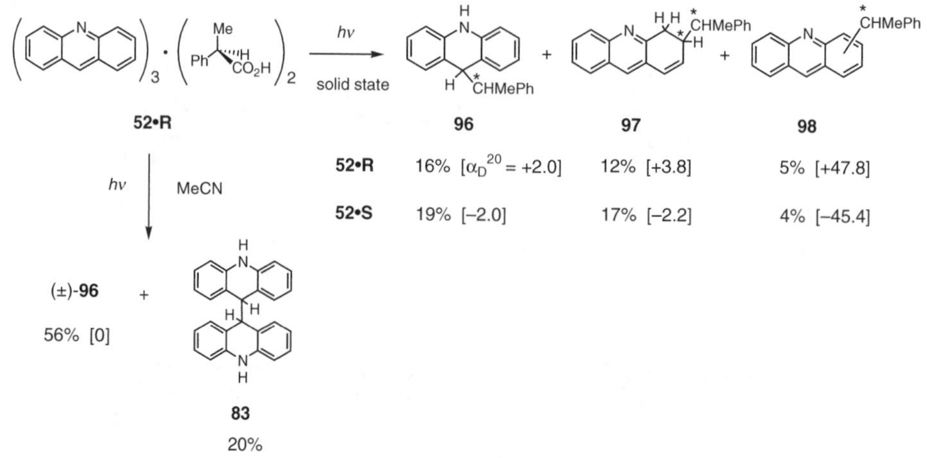

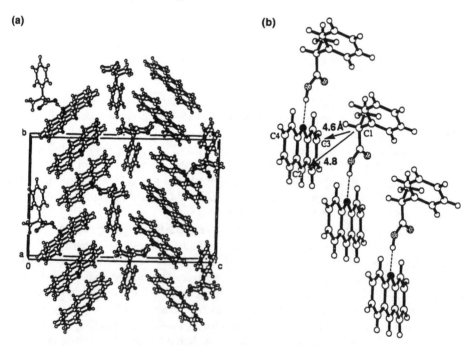

Scheme 8.8 Photoreaction of acridine and (*R*)-(−)- or (*S*)-(+)-phenylpropionic acid in the crystalline state and in solution.

(a)

(b)

Figure 8.6 (a) Molecular arrangement and (b) radical coupling path in the chiral crystal **52•R**. Black and dotted atoms represent nitrogen and oxygen atoms, respectively. *Source*: from ref. 59 with permission. © 1997 Elsevier Science.

•CHMePh is retained to some extent due to the resticted motion possible in the crystal lattice leading to the formation of optically active **96** (Figure 8.6b). The frontal C1—C3 attack by the reactive •CHMePh radical across a distance of 4.6 Å should introduce another asymmetric carbon at the coupling position in **97**, which can be a diastereomeric compound having two chiral carbon centers. The C1—C4 coupling with a distance of 6.1 Å gives another diastereomer of **97** in much smaller yield.

In contrast, the formation of optically inactive product **96** during solution photolysis of **52** and **R** suggests that the original radical species •CHMePh completely racemized in the solution before coupling with the hydroacridane radical between C1 and C2.

8.5. *Photoreaction of α-hydroxycarboxylic acids with aza aromatic compounds*

α-Hydroxycarboxylic acids caused photodecarboxylation and successive dehydrogenation in their two-component crystals with aza aromatic compounds to give selectively ketone or aldehyde products [60]. The series of two-component crystals incorporate α-hydroxycarboxylic acids such as benzilic acid **BE** or (*S*)-(+)-mandelic acid **MA**, and aza aromatic compounds such as acridine **52**, phenanthridine **77** or phenazine **99**. Crystallization from solutions of both components gave the corresponding crystals (**Scheme 8.9**). Two kinds of crystals of phenazine and benzilic acid were obtained simultaneously, **99•BE** (1:2) and **99•BE•H$_2$O** (1:1:1).

Irradiation of the two-component crystals caused decarboxylative dehydrogenation giving benzophenone **1** or benzaldehyde **100** as the main product (**Scheme 8.10** and Table 8.4). Biacridane **83** or biphenanthridane **84** was the minor product. After irradiation of the crystals containing phenazine **99•BE**, **99•BE•H$_2$O** or **99•MA**, the crystal color changed to dark green, indicating formation of the crystalline CT complex **101** of phenazine and dihydrophenazine. However, it was not possible to measure the yield of

Table 8.4 Photoreaction of α-hydroxycarboxylic acids and aza aromatic compounds in the crystalline state and in solution

Components	Conversion (%)		Yield (%)	
	N-Ar	acid	ketone	HNAr–)$_2$
Crystalline state				
52•BE	36	40	48 (**1**)	9 (**83**)
77•BE	41	42	55 (**1**)	24 (**84**)
99•BE	16	42	63 (**1**)	–
99•BE•H$_2$O	19	100	96 (**1**)	–
52•MA	42	56	34 (**100**)	10 (**83**)
77•MA	40	51	38 (**100**)	5 (**84**)
99•MA	12	61	33 (**100**)	–
In MeCN				
52+BE	98	73	86 (**1**)	100 (**83**)
77+BE	91	78	56 (**1**)	92 (**84**)
99+BE	57	90	84 (**1**)	74 (**101**)
52+MA	100	68	26 (**100**)	98 (**83**)
77+MA	57	45	16 (**100**)	98 (**84**)
99+MA	44	90	10 (**100**)	82 (**101**)

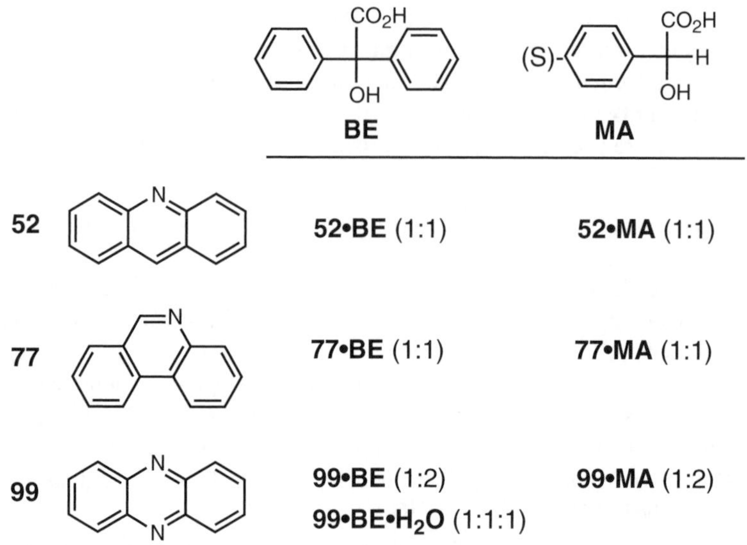

Scheme 8.9 Two-component crystals from aza aromatic compounds and α-hydroxycarboxylic acids.

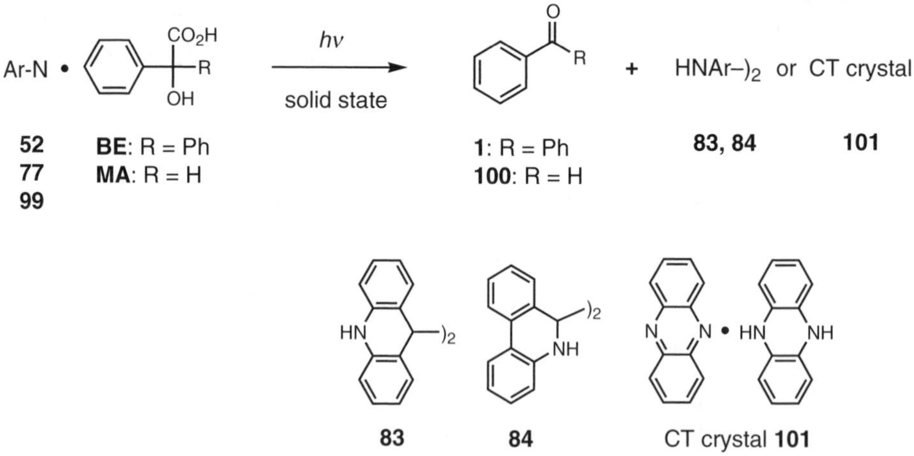

Scheme 8.10 Photoreaction in the two-component crystals of aza aromatic compounds and α-hydroxycarboxylic acids.

101 due to aerial oxidation of dihydrophenazine to phenazine when the irradiated sample was worked up for HPLC analysis.

By way of comparison, photolysis of a mixture of the α-hydroxycarboxylic acid and aza aromatic compounds in acetonitrile under argon caused similar decarboxylation and dehydrogenation to afford two main products, **1** or **100**, and the corresponding dimer **83** or **84** or the dark blue CT crystal **101** as the precipitate. The high yields of **83** and **84** result from the high mobility of the intermediate radical species in the solution. Fairly good yields of benzophenone **1** were obtained with every aza aromatic compound, while the isolated yields of benzaldehyde **100** were low due to its volatility during the separation processes.

Scheme 8.11 shows the possible reaction mechanism. Irradiation of crystal causes electron transfer from acid to the aza aromatic compound followed by proton transfer (or conversely proton transfer followed by electron transfer) to afford α-carboxylate radical (**102**), and hydroacridine radical (**86**), hydrophenanthridine radical (**87**) or phenazine semiquinone radical (**104**). Next, decarboxylation of **102** produces ketyl radical (**105**) followed by dehydrogenation resulting in formation of **1** or **100**. Radical coupling of **86** or **87** also occurs to give the dimer **83** or **84**, while in the crystals with phenazine, disproportionation of semiquinone radical **104** into **99** and dihydrophenazine, followed by CT complexation, affords **101**.

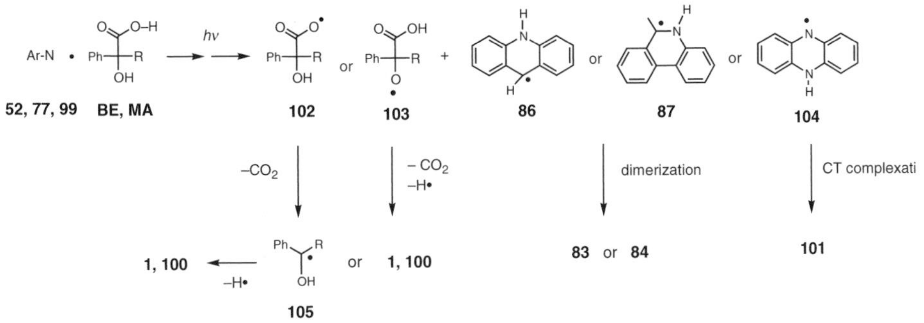

Scheme 8.11 Possible photochemical reaction paths in the crystals of α-hydroxycarboxylic acids and aza aromatic compounds.

The reaction paths at the initial stage can be discussed on the basis of the distances obtained from the crystallographic data as follows. For instance, in the crystal **52•BE**, the $C-O^- —^+H-N$ salt structures are initially formed (Figure 8.7):hence irradiation causes electron transfer from **BE** to **52** through the salt bridges of A pair (1.50 Å) and pair B (1.64 Å) to give the radical species **102** and **86**. Next, decarboxylation of **102** followed by dehydrogenation of **105** as highest priority gives **1** as the main product. Dimerization of **86** occurs to a lesser extent due to the longer coupling distance of 4.96 Å (C1—C2) and 5.33 Å (C1—C3) in the column A, or 5.22 Å (C1—C2) and 5.43 Å (C1—C3) in the column B to give **83** as the minor product.

In the crystal **99•BE** (Figure 8.8), the α-hydroxy group of **BE** is connected to the N atom of **99** to form α-O-H•••N hydrogen bonding, which is different from the CO_2H•••N hydrogen bonding observed in the other crystals with benzilic acid. Therefore, upon irradiation of **99•BE** an electron transfer may occur from **BE** to **99** through the α-O-H•••N hydrogen bonding (2.04 Å) followed by proton transfer to give the α-oxycarboxylic acid radical **103** and phenazine semiquinone radical **104**. Next,

247

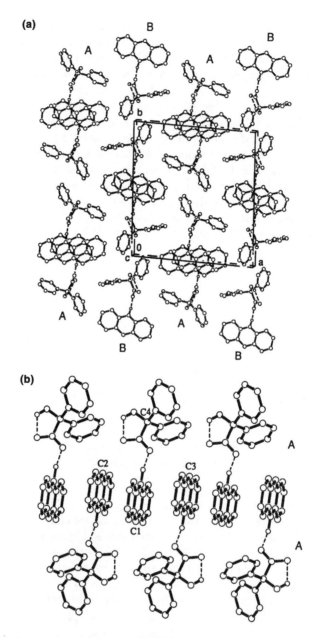

Figure 8.7 Molecular arrangement of the crystal **52•BE** formed from acridine and benzilic acid. *Source*: from ref. 60 with permission. © 1999 Elsevier Science.

cleavage of **103** gives **1** as the main product and also •CO$_2$H radical, which is immediately decomposed to CO$_2$ and H•. Furthermore, disproportionation of **104** into **99** and dihydrophenazine, followed by CT complexation, results in the formation of crystalline CT complex **101**.

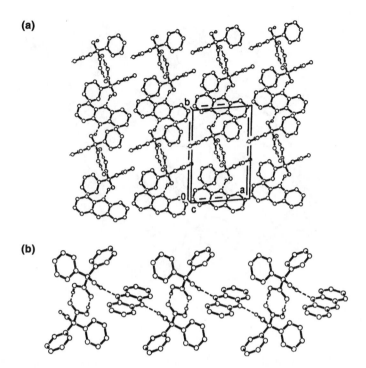

Figure 8.8 Molecular arrangement of the crystal **99•BE** from phenazine and benzilic acid. *Source*: from ref. 60 with permission. © 1999 Elsevier Science.

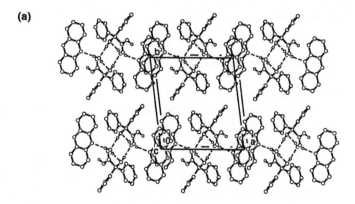

Figure 8.9 Molecular arrangement of the crystal **52•BE•H₂O** formed from phenazine, benzilic acid and water. *Source*: from ref. 60 with permission. © 1999 Elsevier Science.

On the other hand, the three-component crystal **99•BE•H₂O** has the usual $CO_2H \cdots N$ hydrogen bonding between **99** and **BE** (Figure 8.9b). Therefore, photoinduced electron transfer and proton transfer can occur more smoothly through the $CO_2H \cdots N$ hydrogen bonding (1.85 Å) than through the α-O–H$\cdots$N hydrogen bonding in **99•BE**. One of the

(b)

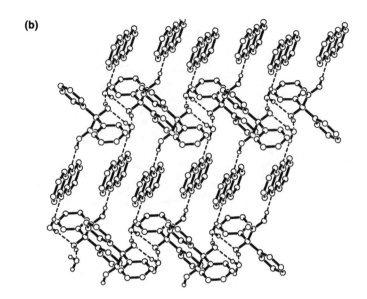

Figure 8.9 Continued.

reasons why the very high yield (96%) of **1** was obtained through reaction of **99•BE•H₂O** is its clean reaction without significant side reactions (as confirmed by HPLC analysis). The parallel stacking structure of **99** with the relatively short plane-to-plane distance of 4.6 Å (Figure 8.9b) may also contribute to the smooth disproportionation of dihydrophenazine, followed by CT complexation to **101**.

9. Absolute Asymmetric Synthesis

Chiral molecules necessarily form chiral crystals. However, even if a molecule is achiral, chiral crystals can form by chiral crystallization [64,65]. Absolute asymmetric synthesis can be achieved by solid-state photoreaction of such chiral crystals because the initial chiral environment in the crystal lattice is retained during the reaction process due to the low mobility of molecules in the crystalline state. This causes enantioselective reaction. This means transformation of the crystal chirality into molecular chirality. Absolute asymmetric synthesis is attractive because it is not necessary to use any chiral source. Furthermore, it is relevant to prebiotic origin of chirality [66]. Therefore one of the topics in organic solid-state chemistry is the promising methodology of absolute asymmetric synthesis [4–12,65,67–70].

In order to achieve absolute asymmetric synthesis, it is necessary first to prepare chiral crystals as reactants by spontaneous crystallization of achiral molecules. Formation of chiral crystals can be confirmed by knowing whether or not the crystals belong to chiral space groups. There are 65 chiral space groups among the 230 possible space groups. In the case of one-component crystals, large numbers of chiral crystals from achiral molecules are already known and compiled in the Cambridge Structural Database and the JCPDS crystal data series. The probability of occurrence of chiral crystallization of achiral molecules is statistically around 5%, and the most frequent chiral space groups are $P2_12_12_1$ and $P2_1$. However, chiral two-component molecular crystals formed from two achiral compounds are scarcely known, except for certain types of CT complexes [71] and a series of helical-type of crystals [72–74].

In fact, about twenty absolute asymmetric syntheses have been achieved by utilizing one-component crystals as the reactants. Most of the successful examples are based on unimolecular (intramolecular) reactions such as di-π-methane photorearrangement of dibenzobarrelenes [75] and other compounds [76], the Norrish type-II photocyclization of oxoamides to β-lactams [77,78], and the photocyclization of imides and thioamides [79]. The bimolecular (intermolecular) reaction is limited to [2 + 2] photocycloaddition in the crystals of 1,4-disubstituted phenylenediacrylate derivatives [80] and 4-[2-pyridyl]ethenyl] cinnamates [81].

In contrast, only three successful examples are known where bimolecular (intermolecular) reactions occur in two-component crystals and mixed crystals. The first one is the [2 + 2] photocycloaddition of chiral mixed crystals of butadienes, rather than a strict two-component crystal [82]. The second excellent example also involves [2 + 2] photocycloaddition, this time of a chiral crystalline CT complex of bis[1,2,5]thiadiazolotetracyanoquinodimethane and *o*-divinylbenzene via a single-crystal-to-single-crystal transformation [83]. The third one is our absolute asymmetric photodecarboxylative condensation that occurred in a chiral two-component crystal formed from acridine and diphenylacetic acid [63]. Herein the three absolute asymmetric bimolecular reactions are precisely explained, with particular emphasis on the enantioselective reaction paths based on the enantiomeric molecular arrangements in their crystal lattices.

9.1.   *[2 + 2] Photodimerization in the mixed crystals of butadienes*

1-(2,6-Dichlorophenyl)-4-phenyl-*trans, trans*-1,3-butadiene **106** crystallizes in chiral space group $P2_12_12_1$ with $a = 9.39$, $b = 4.00$, and $c = 35.30$ Å, $Z = 4$ as colorless needles of mp 107 °C. Irradiation of **106** in the solid state at >290 nm causes [2 + 2] cycloaddition to give a single achiral homodimer **107** as a sole product (**Scheme 9.1**). A thiophene analog **108** forms crystals isomorphous to those of **106** as yellow needles of mp 81 °C ($P2_12_12_1$, $a = 9.15$, $b = 3.99$, $c = 34.98$ Å, $Z = 4$). Similar irradiation of **108** affords the topochemically expected dimer **109**.

Compounds **106** and **108** formed mixed crystals (substitutional solid solutions) by cooling of the melts of both components or on crystallization from ethanol solution. A single, large-sized mixed crystal **106•108** was pulverized and irradiated resulting in [2 + 2] photocycloaddition thereby giving the enantiomeric heterodimers **110a** and **110b** as well as the achiral homodimers **107** and **109** (**Scheme 9.1**)[82]. As expected, some crystals gave predominantly left-handed **110** and others right-handed **110**.

Selective excitation of the thiophene moieties, which absorb longer wavelength light than the phenyl groups, led to a decrease in the formation of achiral homodimers **107** and **109**. In fact, irradiation at longer wavelength (>400 nm) of the homochiral mixed

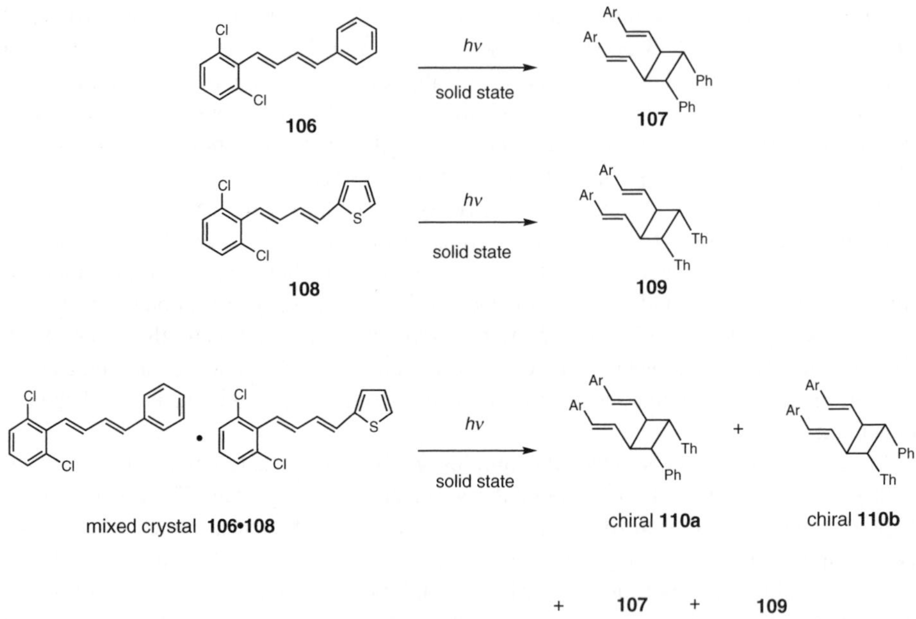

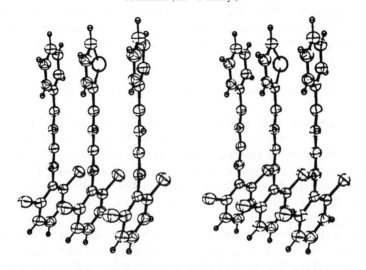

Scheme 9.1 Absolute asymmetric [2 + 2] photocycloaddition in the mixed crystal of butadienes (Th = 2-thienyl).

Figure 9.1 Stereoview of the intermolecular overlap in the mixed crystal of butadienes **106•108**. *Source*: from ref. 67 with permission. © 1979 Am. Chem. Soc.

crystals containing 15% of **108** and 85% of **106** gave the optically active heterodimer **110** with the relatively high optical yield of 70% ee. In contrast, irradiation at shorter wavelength (<350 nm) resulted in lower optical yields as expected due to excitation of the phenyl moieties.

Figure 9.1 shows a stereoview of the mixed crystal containing **106** and **108** in the ratio of 72:28 [67]. The thiophene ring exhibits partial disorder; 40% of the molecules have the conformation shown, while in the remaining 60% the thiophene ring is rotated by 180°. Along the 4-Å translation axis in the mixed crystal the intermolecular olefinic contacts between all pairs of molecules are almost identical. Hence, if there are no unimolecular deformations on excitation, the same probability of cycloaddition in the upward or downward directions, should give equal amounts of enantiomers of **110a** and **110b** (a racemic mixture). However, the experimental result achieving up to 70% ee suggests that when the thiophene moiety is excited, some geometrical deformation occurred and the difference in the interaction of photoexcited **108** with its two nearest neighbors leads to the formation of two diastereomeric transition states, giving rise to the observed asymmetric synthesis.

9.2. [2 + 2] Photocycloaddition in a CT crystal via single-crystal-to-single-crystal transformation

A series of 1:1 CT crystals of bis[1,2,5]thiadiazolotetracyanoquinodimethane (BTDA) **111** as an electron acceptor was obtained by mixing with neat arylolefins such as styrene, *o*-, *m*- or *p*-divinylbenzene as electron donors. X-ray crystallographic analyses of the CT crystals revealed the chiral nature of the red crystals formed between **111** and *o*-divinylbenzene **oDV** due to their belonging to space group $P2_1$, while the other cases were achiral [83].

A water suspension of the crystalline powders of chiral **111•oDV** was irradiated at >540 nm (CT excitation) to cause smooth [2 + 2] cycloaddition and formation of adduct **112** in 84% isolated yield after 15 min at 15 °C (**Scheme 9.2**). Further, the yield increased to 91% by irradiation for 1 h. However, the product **112** formed in the solid-state photoreaction had no optical rotatory power because the reactant CT powder crystals prepared by mixing **111** and neat **oDV** were a racemic mixture of left- and right-handed crystals. Next, single crystals of **111•oDV** were irradiated piece-by-piece

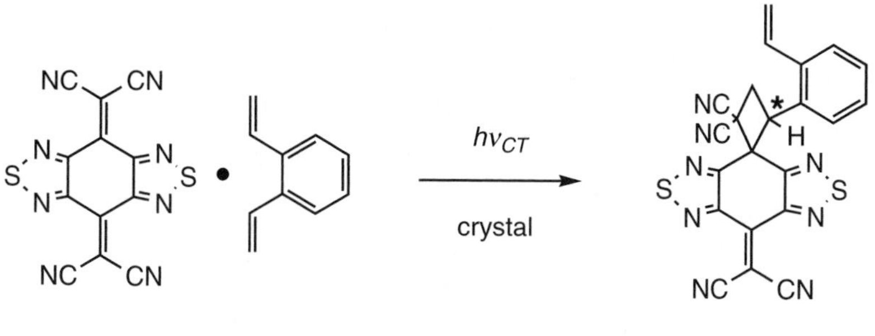

111•oDV

chiral **112**

Scheme 9.2 [2 + 2] Photocycloaddition in the CT crystal of BTDA and *o*-divinylbenzene.

to afford the enantiomeric adduct **112** in an optical purity of 69–73% ee (optical rotation either positive or negative) thereby achieving success in absolute asymmetric synthesis. The ee values increased with lowering of the irradiation temperature; 62%, 71%, 83%, and 95% ee at 40 °C, 15 °C, −40 °C, and −70 °C, respectively.

In contrast, solution photolysis of the CT complex of BTDA **111** and **oDV** in acetonitrile at >450 nm (CT excitation) for 5 h gave the racemic mixture of adduct **112** in only 9% yield, showing that the apparent reactivity of the solid state is much higher than that in solution.

During irradiation of the single crystals of **111•oDV**, the color of the crystals gradually changed from deep red to pale yellow as the photoreaction proceeded. However, all the samples kept a clear appearance and never became opaque, suggesting that the reaction proceeds via single-crystal-to-single-crystal transformation. After many trials of irradiation of the single crystals of **111•oDV** in glass capillaries, X-ray crystallographic analysis of the as-prepared adduct **112** was successful. Comparison of the crystal data in Table 9.1 indicates that the as-prepared adduct **112** is almost isomorphous with the reactant **111•oDV**, confirming that the single-crystal-to-single-crystal reaction proceeded without decomposition of the initial crystal structure.

Figure 9.2 shows the molecular arrangements in the crystals before (a, b) and after irradiation (c, d), both of which are very similar. In the crystal lattice of reactant **111•oDV**, linear ribbon networks are formed by S•••N≡C weak contacts of BTDA **111** (3.14 and 3.43 Å) along the *a*-axis. The **oDV** molecules are incorporated in the cavity formed between two ribbons as a noncentrosymmetric arrangement on the *ab* face. This type of molecular arrangement is quite common in CT crystals of **111** and aromatic hydrocarbons [84]. Perpendicular to this arrangement, a one-dimensional columnar structure is formed along the *c*-axis, in which two types of molecular overlaps (type 1 and type 2) are repeated alternately. Two olefinic groups are aligned nearly in parallel in both types of overlap. The interatomic distances are 3.44 Å (C5—C29) and 3.87 Å (C7—C30) for type 1, and 3.75 Å (C5—C29) and 3.55 Å (C7—C30) for type 2. These distances are short enough to permit topochemical [2 + 2] cycloaddition in the solid state under the 4 Å rule (Figure 9.3a and b).

Table 9.1 Crystal data of the reactant crystal **111•oDV** and its product **112**

	111•oDV	**112**
Formula	$C_{22}H_{10}N_8S_2$	$C_{22}H_{10}N_8S_2$
Space group	$P2_1$	$P2_1$
a(Å)	8.866(2)	9.120(1)
b(Å)	16.764(3)	16.394(2)
c(Å)	7.006(3)	7.181(1)
β (deg)	99.63(2)	106.27(1)
Z	2	2
D_{calc}(g cm^{-3})	1.457	1.452
R	0.042	0.082

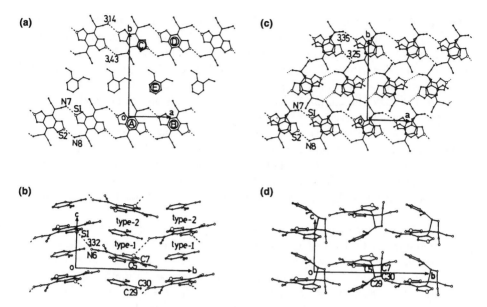

Figure 9.2 Molecular arrangements in the reactant **111•oDV** along (a) the *c*-axis and (b) the *a*-axis. Molecular arrangements in the product **112** along (c) the *c*-axis and (d) the *a*-axis. *Source*: from ref. 83 with permission. © 1994 Am. Chem. Soc.

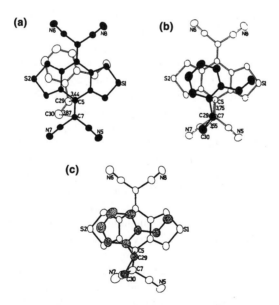

Figure 9.3 (a) Type 1 and (b) type 2 molecular overlaps in the reactant **111•oDV**. The interplanar distances and dihedral angles are 3.40 Å and 2.6° in type 1 and 3.44 Å and 2.6° in type 2. (c) Molecular structure of the chiral product **112**. The bond lengths of C(5)–C(29) and C(7)–C(30) are 1.66(2) and 1.55(2) Å. *Source*: from ref. 83 with permission. © 1994 Am. Chem. Soc.

On the other hand, the striking resemblance of molecular arrangements of the chiral product **112** in Figure 9.2c and d suggests that the cycloaddition occurred without a significant change in the crystal structure of the reactant **111•oDV**. Only slight movement of the reactive sites along the stacking c-axis enables the [2 + 2] cycloaddition of **111** and **oDV** and allows the topochemical and single-crystal-to-single-crystal transformation of **111•oDV** into chiral **112**. But a question still remains as to which of the type 1 or type 2 molecular overlap is favorable in the topochemical [2 + 2] cycloaddition. Direct evidence that the cycloaddition proceeded predominantly via the type 2 overlap in **111•oDV** could be obtained by comparison of the molecular arrangements in Figure 9.3b and c. This experimental observation suggests that the stronger CT interaction through the face-to-face overlap plays a more important role than the closest atomic contact between the reactive sites for this case of cycloaddition induced by CT excitation.

Finally, the absolute structure of the selected single crystal of **111•oDV** was successfully determined by the anomalous dispersion of the sulfur atom using CuKα radiation during the X-ray crystallographic analysis. Irradiation of the same sample afforded (+)-**112**. The absolute configuration of **112** could not be determined by the X-ray anomalous dispersion, but it could be deduced as (R)-(+)-**112** by considering the predominant cycloaddition via the type 2 overlap in **111•oDV**. Hence, X-ray crystallographic analysis can be a powerful tool for the direct elucidation of reaction pathways in the single-crystal-to-single-crystal transformations.

9.3. Absolute asymmetric photodecarboxylative condensation in a two-component crystal of acridine and diphenylacetic acid

Despite acridine and diphenylacetic acid being achiral molecules, it was found that these two compounds formed a chiral two-component crystal on spontaneous crystallization, and further irradiation of this chiral crystal caused photodecarboxylation and then enantioselective radical coupling to give a chiral condensation product [63]. This kind of absolute asymmetric synthesis is interesting and suggestive because the solid-state photodecarboxylation is not a topochemical reaction, but is accompanied by gradual decomposition of the initial crystal structure as the reaction proceeds, which is different from the two examples of [2 + 2] photocycloadditions described above.

The chiral two-component crystal **52•DPA** was prepared by crystallization from a 1:1 solution of acridine **52** and diphenylacetic acid **DPA** in acetonitrile (**Scheme 9.3**). The crystal exists as light-yellow rods and its melting point (101 °C) is lower than those of either **52** (107 °C) or **DPA** (148 °C). The crystal belongs to chiral space group $P2_12_12_1$ with the cell parameters $a = 12.254$, $b = 7.226$, $c = 12.737$ Å, $Z = 4$, $D_c = 1.250 \, \text{g cm}^{-3}$, $R = 0.036$. Both enantiomorphous crystals of the M- and P-form were obtained by spontaneous crystallization from the solution, and the handedness was easily differentiated by the measurement of solid-state circular dichroism (CD) spectra as Nujol mulls (Figure 9.4). The two CD curves show an excellent mirror-image relationship.

The absolute structure of reactant **52•DPA**, i.e. whether a given crystal has the M- or P-form, was determined by X-ray anomalous dispersion of the oxygen atom despite its

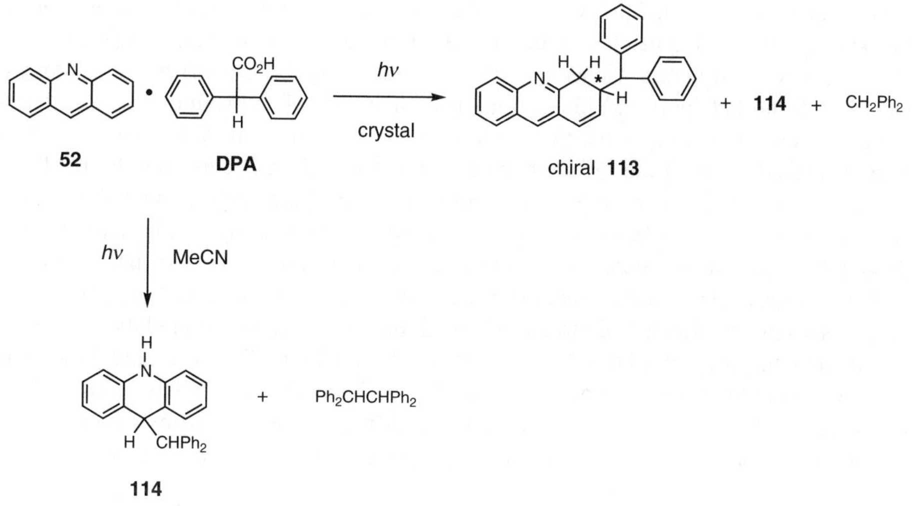

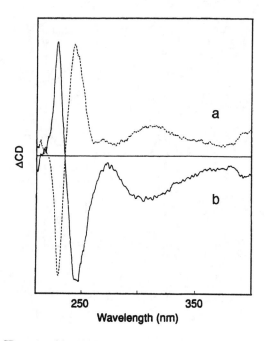

Scheme 9.3   Absolute asymmetric photodecarboxylative condensation in the two-component crystal of acridine and diphenylacetic acid.

Figure 9.4   Solid-state CD spectra of the (a) *M*-crystal and (b) *P*-crystal formed from acridine and diphenylacetic acid. *Source*: from ref. 85 with permission. © 1997 Am. Chem. Soc.

small value ($\Delta f'' = 0.032$ for CuKα radiation), because even slight modification of **52·DPA** by introducing heavier atoms such as sulfur and chlorine may cause changes in the initial crystal structure. Hence a piece of single crystal of **52·DPA** was cut into two pieces and one half polished into a spherical shape and the absolute structure determined as being the *M*-crystal with a high degree of certainty. The solid-state CD spectrum measured by using the second half corresponded to the CD curve a in Figure 9.4.

Figure 9.5 illustrates the molecular arrangement in the *M*-crystal with the correct absolute structure. N•••H–O hydrogen bonding (1.83 Å) between the acridine and diphenylacetic acid molecules is present in the crystal lattice. The two phenyl planes and the carboxyl plane of the diphenylacetic acid molecule in the molecular pair form a propeller-like conformation of three blades with the torsion angles around the C–C bond to the methine group of 13.0° (Ph-1), 50.3° (Ph-2) and 33.9° (COO) (Figure 9.6a). The helicity around the C–H bond of the methine group is counterclockwise (minus), which is termed as the *M*-crystal. The crystal chirality is generated from the existence of molecular pairs of counterclockwise propellers alone in the unit cell ($Z = 4$).

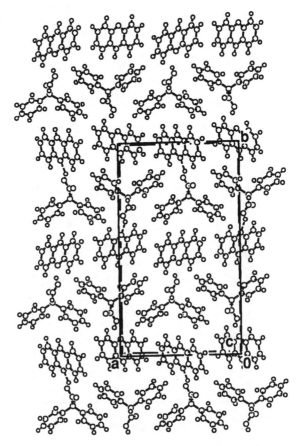

Figure 9.5 Molecular arrangement in the *M*-crystal formed from acridine and diphenylacetic acid. *Source*: from ref. 63 with permission. © 1996 Am. Chem. Soc.

Conversely, only the molecular pairs with opposite torsion angles of $-13.0°$ (Ph-1), $-50.3°$ (Ph-2) and $-33.9°$ (COO) are observed in the *P*-crystal (Figure 9.6b).

The *M*-crystals of **52•DPA** prepared by seeding were pulverized and irradiated with a high-pressure mercury lamp through Pyrex glass (>290 nm irradiation) under argon for 3 h at 15 °C on a preparative scale. Solid-state photodecarboxylation and enantioselective condensation occurred to give chiral condensation product (–)-**113** as the main product with $[\alpha]_D^{20} = -30$ in 35% ee and in 37% chemical yield, hence achieving absolute asymmetric synthesis (**Scheme 9.3**). An achiral condensation product **113** and decarboxylated product diphenylmethane **114** were also obtained as minor products. Conversely, irradiation of *P*-crystals resulted in the formation of the opposite-handed condensation product (+)-**113** with $[\alpha]_D^{20} = +30$ in 33% ee and in 38% chemical yield.

In contrast, solution-phase photolysis of acridine **52** and diphenylacetic acid **DPA** in acetonitrile did not produce chiral product **113**, but rather gave the achiral condensation product **114** and 1,1,2,2-tetraphenylethane in 74% and 24% yield, respectively, at complete conversion of **DPA** (**Scheme 9.3**).

It is most important for absolute asymmetric synthesis using the solid-state reactions to determine the absolute configurations of both the reactant and product. Therefore the absolute configuration of the product was determined as follows. The pure enantiomer (–)-**113** was separated by preparative HPLC using a chiral column and a sufficiently

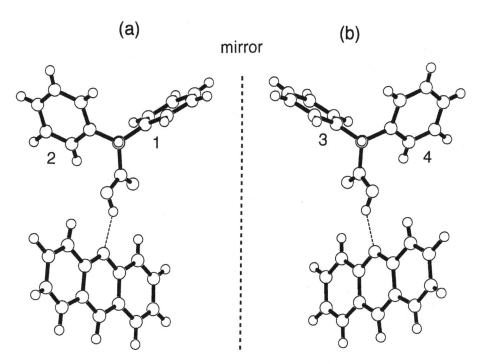

Figure 9.6 Molecular pairs arranged in the (a) *M*-crystal and (b) *P*-crystal formed from acridine and diphenylacetic acid.

heavy sulfur atom ($\Delta f'' = 0.557$ for CuKα radiation) was introduced by preparing the trifluoromethanesulfonate salt of methylated (–)-**113** CF$_3$SO$_3^-$•$^+$CH$_3$-(–)-**113**, which did not affect the configuration of the asymmetric carbon. Thus the absolute configuration was determined as (S)-CF$_3$SO$_3^-$•$^+$CH$_3$-(–)-**113** with a high degree of certainty, based on the anomalous dispersion of the sulfur atom with CuKα radiation. This means that the *M*-crystal produces predominantly (S)-(–)-**113** in the solid-state reaction, and conversely the *P*-crystal gives (R)-(+)-**113**.

The solid-state reactivity was further examined by irradiating single crystals of *M*- and *P*-form piece-by-piece followed by HPLC analysis (Table 9.2). Prolonged irradiation led to a low yield of the chiral product **113** probably due to partial decomposition, but the ee value was almost unchanged around 35% ee (entries 2 and 3). The photoreactivity was retained even at $-70\,^\circ$C with an accompanying slight decrease in reaction rate and yield of the chiral product, but the ee value remained unchanged (entry 5). The reaction was effected not only using UV (290–390 nm) (entries 1–3, 5 and 6) but also visible light (>390 nm) (entries 4 and 7). The crystal has a broad absorption band over the wavelength range 200–450 nm, which almost equals the sum of those of the components acridine (200–450 nm) and diphenylacetic acid (200–300 nm). Therefore this reaction is induced through light absorption by the acridine molecule at >390 nm, and it is the acridine molecule which acts as the excited species in the crystal.

The enantioselective reaction mechanism can be explained by considering the photochemical aspects and also the molecular arrangement (**Scheme 9.4** and Figure 9.7). Irradiation of the crystal excites the acridine component followed by electron transfer from diphenylacetic acid to acridine to give a cation radical (**116**) and an anion radical (**115**). Then proton transfer affords the diphenylacetate radical (**117**) and hydroacridine radical species (**118–121**). The charge densities on the acridine anion radical **115** calculated by semiempirical PM3 in vacuum are localized on N1 (-0.94), C20 and C27 (-0.44) (for numbering, see Figure 9.7). On the other hand, the short distances of H2---N1, H2---C20, H2---C27, and H2---C28, are estimated from the crystallographic data as 1.94, 3.13, 2.94, and 3.31 Å, respectively. Other distances are longer than 4 Å. Both higher charge density and shorter distance are required for proton transfer in the

Table 9.2   Photoreaction of the single crystals of *M*-and *P*-**52•DPA**

Entry	Crystal 52•DPA	Irradiation			Conversion (%)		Yield^a (%)			ee (%)
		°C	h	nm	52	DPA	chiral 113	114	CH$_2$Ph$_2$	113
1	*M*	10	1	290–390	28	45	50	2	11	36 (S)
2	*M*	−30	0.5	290–390	6	13	58	2	8	35 (S)
3	*M*	−30	5	290–390	52	71	28	2	6	33 (S)
4	*M*	−30	1	>390	39	63	39	2	9	36 (S)
5	*M*	−70	1	290–390	12	23	34	2	4	39 (S)
6	*P*	−30	1	290–390	23	35	51	4	9	37 (R)
7	*P*	−30	1	>390	31	51	32	2	7	35 (R)

^{a} Yield based on consumed **DPA**.

crystal lattice, which is a different situation from that in solution. Hydroacridine radical species **118** and **119**, which satisfy both such conditions, should be preferably produced with their higher stability. Next decarboxylation of **117** gives the diphenylmethyl radical **122**. In the chiral crystal lattice of *M*-**52•DPA**, the molecular pairs of acridine and diphenyl acetic acid stack along the *c*-axis with a interplanar distance of 5.46 Å to form a column structure (Figure 9.7). Radical coupling between •C1 of **122** and •C29 of **119**

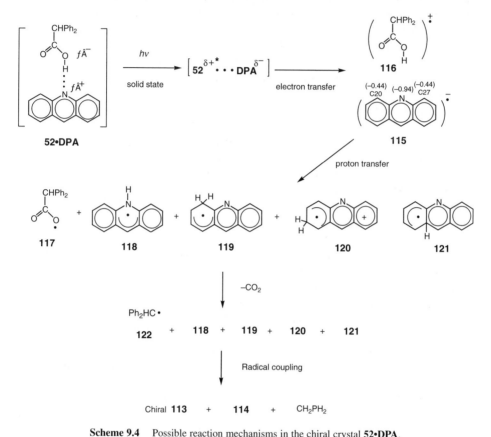

Scheme 9.4   Possible reaction mechanisms in the chiral crystal **52•DPA**.

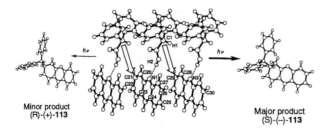

Figure 9.7   Enantioselective radical coupling path in the crystal *M*-**52•DPA**. *Source*: from ref. 63 with permission. © 1996 Am. Chem. Soc.

over the shortest distance of 5.11 Å can occur with highest priority to give (*S*)-(–)-**113** condensation product as the major enantiomer. Radical coupling of •C1 and •C26 over a distance of 6.52 Å also produces (*S*)-(–)-**113**. Conversely, the (*R*)-(+)-product as a minor enantiomer is formed by the radical coupling between •C1 of **122** and •C21 or •C30 of **119** over the longer distances of 6.79 and 8.97 Å, respectively. The radical coupling ratio giving S:R is calculated to be approximately 2:1 from the experimental ee value of around 35% ee. The radical coupling is necessarily accompanied by slightly larger movement of the radical species in the crystal lattice, in contrast to the well-known topochemical [2 + 2] photocycloaddition, for which a distance of less than 4 Å is indispensable [2]. Such decarboxylation and condensation reactions lead to the gradual decomposition of the crystal structure of the reactant. Nevertheless, it is interesting that almost constant ee values of 35% were obtained over a wide range of conversions of acridine (6–52%) and diphenylacetic acid (13–71%) (entries 2, 3 in Table 9.2).

The achiral condensation product **114** is scarcely produced in the solid-state photoreaction (Table 9.2), most probably due to the steric hindrance of the bulky diphenylmethyl radical, which is different from the solution reaction which gives **114** in high yield. In addition, formation of diphenylmethane and regeneration of acridine **52** are caused by hydrogen abstraction by the diphenylmethyl radical **122** from the hydroacridine radical **118**, because the •C1–H2 distance (3.11 Å) between these radicals is short enough to allow abstraction of the hydrogen H2. The regeneration of acridine is the reason for the lower conversion of acridine compared to diphenylacetic acid (Table 9.2).

This success in absolute asymmetric synthesis led us to prepare other propeller-type chiral crystals. Combination of diphenylacetic acid **DPA** with various aza aromatic compounds such as phenanthridine **77**, benzo[*f*]quinoline **123**, quinoline **124** or phenazine **99** resulted in the discovery of a new chiral crystal **77•DPA**, belonging to space group $P2_1$ [85]. The other crystals **123•DPA**, **124•DPA**, and **99•DPA** were achiral, in space groups are $P2_1/a$, $P2_1/c$, and $P2_1/n$, respectively.

The solid-state CD spectra of the *M*- and *P*-crystals of **77•DPA** obtained by spontaneous crystallization from methanol solution are shown in Figure 9.8. The absolute structure of *P*-**77•DPA** was determined by the same method as that of **52•DPA** (as described above, based on the X-ray anomalous dispersion of the oxygen atom). The crystal chirality of *P*-**77•DPA** is also generated from the propeller-like structure of the two phenyl planes and the carboxyl plane; the torsion angles around the C–C bond to the methine group are −13.0° (Ph-3), −50.3° (Ph-4), and −33.9° (COO) (Figure 9.9b). Similarly, the molecular pairs with opposite torsion angles of 13.0° (Ph-1), 50.3° (Ph-2), and 33.9° (COO) are present in the crystal *M*-**77•DPA** (Figure 9.9a). The two kinds of molecular pairs (a and b) have a mirror image relationship to each other.

It is understandable that the diphenylacetic acid molecule is achiral in the solution phase due to the free rotation of flexible planes, but frozen into a chiral conformation in the crystals of **52•DPA** and **77•DPA**. Hence, this type of chiral crystallization is like a spontaneous resolution of racemic compounds. It is known that the chiral crystal of benzophenone (space group $P2_12_12_1$) exhibits a similar situation [86]. The two phenyl planes of benzophenone molecules have distinct torsion angles in the crystal. Although two torsional conformations with a mirror image relationship are possible, only the molecules of one single absolute configuration are present in a given crystal.

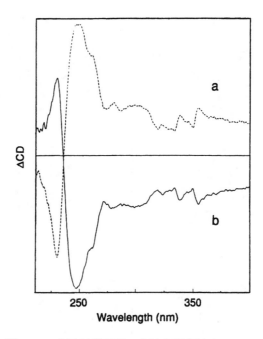

Figure 9.8   Solid-state CD spectra of (a) *M*-**77•DPA** and (b) *P*-**77•DPA** formed from phenanthridine **77** and diphenylacetic acid **DPA**. *Source*: from ref. 85 with permission. © 1997 Am. Chem. Soc.

However, such a chiral conformation cannot necessarily induce chiral crystallization. The crystal of benzo[*f*]quinoline **123** and **DPA** crystallizes in an achiral space group ($P2_1/a$, $Z = 4$) where two molecular pairs of the left-handed conformation and two of the right-handed one coexist in each unit cell (Figure 9.9c, d). Therefore, the chirality is nullified and an achiral crystal results. In the case of the achiral crystal **99•DPA** ($P2_1/n$), two antipodal molecules of **DPA** connect to two N atoms of a phenazine molecule to form a 2:1 hydrogen bonding pair, whereby an inversion center exists in the center of the phenazine plane, resulting in loss of crystal chirality (Figure 9.9e).

The prediction whether the molecular pairs will be self-assembled in only a one-handed conformation or in both-handed conformations in spontaneous crystallization still remains difficult, since it depends on only a small difference among the molecular structures of the aza aromatic compounds. Nevertheless, a principal guideline has been established that the torsional conformation of a group such as a phenyl plane is effective for induction of chirality in the formation of two-component crystals.

Finally, solid-state photoreaction of the chiral crystal **77•DPA** was examined. Irradiation of the pulverized *P*-crystals of **22•10** with a high-pressure mercury lamp under argon caused photodecarboxylation and then condensation to give an achiral condensation product **27**. Unfortunately, absolute asymmetric synthesis was not achieved this time, probably due to the occurrence of subsequent dehydrogenation from **125** leading to formation of the achiral product **126** (**Scheme 9.5**).

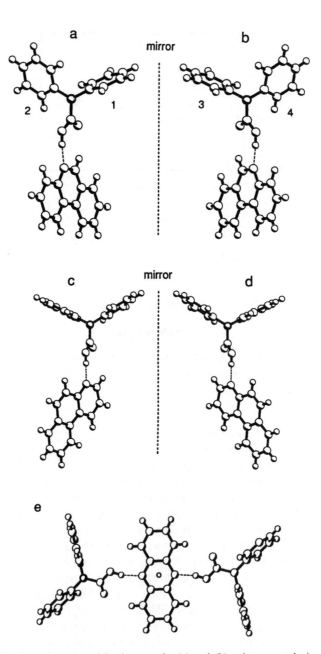

Figure 9.9 Molecular pairs arranged in the crystals. (a) and (b) exist separately in *M*-**77•DPA** and *P*-**77•DPA**. (c) and (d) coexist in achiral **123•DPA**. (e) is a 1:2 symmetrical pair packed in achiral **99•DPA**. *Source*: from ref. 85 with permission. © 1997 Am. Chem. Soc.

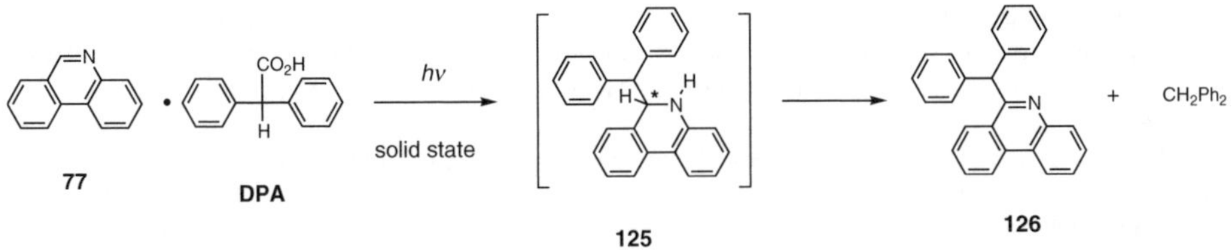

Scheme 9.5   Solid state photoreaction in the chiral two-component crystal formed from phenanthridine and diphenylacetic acid.

10.   Concluding Remarks

Despite the fact that bimolecular photoreactions between two different molecules in two-component crystals are less developed in comparison to a large number of unimolecular photoreactions in one-component crystals, the reaction types are full of variety, as shown in this chapter. Further development of various types of bimolecular photoreactions is desirable to extend the scope of solid-state photochemistry into a versatile field. This may be accomplished not only with two-component but also with multi-component crystals as reactants.

Solid-state photochemistry of two-component crystals is constructed from two steps, the preparation of two-component crystals utilizing weak non-covalent interactions such as hydrogen bonding and charge transfer interactions, and the solid-state photoreaction itself. Two-component crystals have been shown to be useful in (1) induction of photoreactivity *via* photoinduced electron transfer, (2) achievement of higher regioselective, stereoselective and enantioselective reactions than those of solution reactions and specific reactions different from those in solutions, and (3) development of various types of reactions which cannot be expected in one-component crystals. Thus, methods of preparation of two-component crystals are of key importance. In addition, such study of preparation of two- and multi-component crystals will contribute to the development of crystal engineering.

Information on molecular motion in crystals is important because very little motion should occur during reaction despite molecules in crystals being much more restricted than those in solution. Photoreaction cannot occur in the crystal whose molecular packing is too dense. The success of absolute asymmetric decarboxylative condensation by the radical coupling over longer distances of 5.1 and 6.8 Å [63] than the 4 Å of [2 + 2] cycloadditions [2] provides important information about molecular motion in the crystal lattice. At the present time, reaction paths are discussed only based on static crystal data obtained by X-ray structure analysis. Physicochemical studies of dynamical motion of excited species in the crystal lattice will lead to further developments of solid-state photochemistry in the future.

References

1.   M. D. COHEN and G. M. SCHMIDT. *J. Chem. Soc.* (1964) 1996.
2.   G. M. SCHMIDT. *Pure Appl. Chem.* **27** (1971) 647.

3. "Organic solid State Chemistry", edited by G. R. DESIRAJU. (Elsevier, Amsterdam, 1987).
4. "Photochemistry in Solid Surfaces", edited by M. ANPO and T. MATSUURA. (Elsevier, Amsterdam, 1989).
5. "Photochemistry in Organized and Constraint Media", edited by V. RAMAMURTHY. (VCH, Weinheim, 1991).
6. "Reactivity in Molecular Crystals", edited by Y. OHASHI. (VCH, Kodansha, Tokyo, l993).
7. "Supramolecular Photochemistry", edited by V. RAMAMURTHY and K. SCHANZE. (Marcel Decker, New York, 1998).
8. F. TODA. *Acc. Chem. Res.* **28** (1995) 480.
9. J. N. GAMLIN, R. JONES, M. LEIBOVITCH, B. PATRICK, J. R. SCHEFFER and J. TROTTER. *Acc. Chem. Res.* **29** (1996) 203.
10. Y. ITO. *Synthesis* (1998) 1.
11. K. TANAKA and F. TODA. *Chem. Rev.* **100** (2000) 1025.
12. H. KOSHIMA and T. MATSUURA. *Kokagaku* **19** (1995) 10.
13. H. KOSHIMA and T. MATSUURA. *J. Photochem. Photobiol.* **100** (1996) 85.
14. A. I. KITAIGORODSKY. "Molecular Crystals and Molecules", (Academic Press, New York, 1973).
15. G. R. DESIRAJU. "Crystal Engineering: The Design of Organic Solids", (Elsevier, Amsterdam, 1989).
16. G. A. JEFFREY. "An Introduction to Hydrogen Bonding", (Oxford University Press, New York, 1997).
17. H. KOSHIMA, D. P. HESSLER BITTL, G. MIYOSHI, Y. WANG and T. MATSUURA. *J. Photochem. Photobiol. A: Chem.* **86** (1995) 171.
18. H. KOSHIMA, Y. CHISAKA, Y. WANG, X. YAO, H. WANG, R. WANG, A. MAEDA and T. MATSUURA. *Tetrahedron* **50** (1994) 13617.
19. J. B. MENG, W. G. WANG, H. G. WANG, T. MATSUURA, H. KOSHIMA, I. SUGIMOTO and Y. ITO. *Photochem. Photobiol* **57** (1993) 597.
20. H. KOSHIMA, H. ICHIMURA and T. MATSUURA. *Chem. Lett.* (1994) 847.
21. H. KOSHIMA, H. ICHIMURA, K. HIROTSU, I. MIYAHARA, Y. WANG and T. MATSUURA. *J. Photochem. Photobiol. A: Chem.* **85** (1995) 225.
22. H. KOSHIMA and T. MATSUURA. *Mol. Cryst. Liq. Cryst.* **277** (1996) 55.
23. H. KOSHIMA, K. DING, I. MIYAHARA, K. HIROTSU, M. KANZAKI and T. MATSUURA. *Photochem. Photobiol. A: Chem.* **87** (1995) 219.
24. J. B. MENG, D. C. FU, Z. H. GAO, R. J. WANG, H. G. WANG, I. SAITO, R. KASATANI and T. MATSUURA. *Tetrahedron* **46** (1990) 2367.
25. J. B. MENG, W. G. WANG, G. X. XIONG, Y. M. WANG, D. C. FU, D. M. DU, R. J. WANG, H. G. WANG, H. KOSHIMA and T. MATSUURA. *J. Photochem. Photobiol. A: Chem.* **74** (1993) 43.
26. J. B. MENG, D. M. DU, G. X. XIONG, W. G. WANG, Y. M. WANG, H. KOSHIMA and T. MATSUURA. *J. Heterocyclic Chem.* **31** (1994) 121.
27. Y. ITO, J. B. MENG, S. SUZUKI, Y. KUSUNAGA, T. MATSUURA and K. FUKUYAMA. *Tetrahedron Lett.* **26** (1985) 2093.
28. J. B. MENG, Z. L. ZHU, R. J. WANG, X. K. YAO, Y. ITO, H. IHARA and T. MATSUURA. *Chem. Lett.* (1990) 1247.
29. R. P. WAYNE. "Photochemistry", (Butterworths, London, 1970) pp. 147–149.
30. Y. ITO, T. MATSUURA and K. FUKUYAMA. *Tetrahedron Lett.* **29** (1988) 3087.
31. H. KOSHIMA, X. YAO, H. WANG, R. WANG and T. MATSUURA. *Tetrahedron Lett.* **35** (1994) 4801.
32. M. LAHAV, L. LERSEROWITZ, R. POPOVITS-BIRO and C.-P. TANG. *J. Am. Chem. Soc.* **100** (1978) 2542.
33. C. P. TANG, H. C. CHANG, R. POPOVITZ-BIRO, F. FROLOW, M. LAHAV, L. LERSEROWITZ and R. K. MCMULLAN. *J. Am. Chem. Soc.* **107** (1985) 4058.
34. H. C. CHANG, R. POPOVITZ-BIRO, M. LAHAV and L. LERSEROWITZ. *J. Am. Chem. Soc.* **109** (1987) 3883.
35. Y. WEISINGER-LEWIN, M. VAIDA, R. POPOVITZ-BIRO, H. C. CHANG, F. MANNIG, F. FROLOW, M. LAHAV and L. LEISEROWITZ. *Tetrahedron* **43** (1987) 1449.
36. H. KOSHIMA, Y. WANG, T. MATSUURA, I. MIYAHARA, H. MIZUTANI, K. HIROTSU, T. ASAHI and H. MASUHARA. *J. Chem. Soc. Perkin Trans. 2* (1997) 2033.
37. T. ASAHI, Y. MATSUO and H. MASUHARA. *Chem. Phys. Lett.* **256** (1996) 525.
38. Y. ITO, S. ENDO and S. OHBA. *J. Am. Chem. Soc.* **119** (1997) 5974.
39. J. B. MENG, W. G. WANG, Y. M. WANG, H. G. WANG, H. KOSHIMA and T. MATSUURA. *Mol. Cryst. Liq. Cryst.* **242** (1994) 135
40. A. MATSUMOTO, T. ODANI, M. CHIKADA, K. SADA and M. MIYATA. *J. Am. Chem. Soc.* **121** (1999) 11122.

41. A. MATSUMOTO, S. NAGAHAMA and T. ODANI. *J. Am. Chem. Soc* **122** (2000) 9109.
42. A. MATSUMOTO, T. ODANI, K. SADA, M. MIYATA and K. TASHIRO. *Nature* **405** (2000) 328.
43. M. HASEGAWA, Y. SUZUKI, F. SUZUKI and H. NAKANISHI. *J. Polym. Sci. Part A-1* **7** (1969) 743.
44. M. HASAGAWA, M. AOYAMA, Y. MAEKAWA and Y. OHASHI. *Macromolecules* **22** (1989) 1568.
45. M. HASEGAWA, K. KINBARA, Y. ADEGAWA and K. SAIGO. *J. Am. Chem. Soc.* **115** (1993) 3820.
46. (a) J. A. R. P. SARMA and G. R. DESIRAJU. *J. Chem. Soc., Chem. Commun.* (1984) 145. (b) J. A. R. P. SARMA and G. R. DESIRAJU. *J. Am. Chem. Soc.* **108** (1986) 2791. (c) J. A. R. P. SARMA and G. R. DESIRAJU. *J. Chem. Soc., Perkin Trans. 2* (1987) 1187.
47. (a) N. HAGA, H. NAKAJIMA, H. TAKAYANAGI and K. TOKUMARU. *J. Chem. Soc., Chem. Commun.* (1997) 1171. (b) N. HAGA, H. NAKAJIMA, H. TAKAYANAGI and K. TOKUMARU. *J. Org. Chem.* **63** (1998) 5372.
48. Y. ITO, T. MATSUURA and K. FUKUYAMA. *Tetrahedron Lett.* **29** (1988) 2087.
49. B. BOSCH, S. M. HUBIG, S. V. LINDEMAN and J. K. KOCHI. *J. Org. Chem.* **63** (1998) 592.
50. N. HAYAHI, Y. MAZAKI and K. KOBAYASHI. *Tetrahedron Lett.* **35** (1994) 5883.
51. (a) Y. MAZAKI, N. HAYASHI and K. KOBAYASHI. *J. Chem. Soc., Chem. Commun.* (1992) 1381. (b) Y. MAZAKI, K. AWAGA and K. KOBAYASHI. *J. Chem. Soc., Chem. Commun.* (1992) 1661. (c) N. HAYASHI, Y. MAZAKI and K. KOYABASHI. *Chem. Lett.* (1992) 1689.
52. "Photoinduced Electron Transfer, part C", edited by M. A. FOX and M. CHANON. (Elsevier, Amsterdam, 1988).
53. C. BUDAC and P. WAN. *J. Photochem. Photobiol. A: Chem.* **67** (1994) 135.
54. (a) R. NOYORI, M. KATO, M. KAWANISHI and H. NOZAKI. *Tetrahedron* **25** (1969) 1125. (b) D. R. G. BRIMAGE, R. S. DAVIDSON and P. R. STEINER. *J. Chem. Soc., Perkin Trans. 1* (1973) 526. (c) K. OKADA, K. OKUBO and M. ODA. *Tetrahedron Lett.* **30** (1989) 6733. (d) K. OKADA, K. OKUBO and M. ODA. *J. Photochem. Photobiol. A: Chem.* **57** (1991) 265.
55. (a) K. TSUJIMOTO, N. NAKAO and M. OHASHI. *J. Chem. Soc., Chem. Commun.* (1992) 366. (b) J. LIBMAN. *J. Am. Chem. Soc.* **97** (1975) 4139.
56. (a) H. KOSHIMA, K. DING and T. MATSUURA. *J. Chem. Soc., Chem. Commun.* (1994) 2053. (b) H. KOSHIMA, K. DING, Y. CHISAKA, T. MATSUURA, I. MIYAHARA and K. HIROTSU. *J. Am. Chem. Soc.* **119** (1997) 10317.
57. H. KOSHIMA, K. DING, Y. CHISAKA, T. MATSUURA, Y. OHASHI and M. MUKASA. *J. Org. Chem.* **61** (1996) 2352.
58. H. KOSHIMA, K. DING, Y. CHISAKA and T. MATSUURA. *Tetrahedron: Asymmetry* **6** (1995) 101.
59. H. KOSHIMA, T. NAKAGAWA and T. M. MATSUURA. *Tetrahedron Lett.* **38** (1997) 6063.
60. H. KOSHIMA, E. HAYASHI, K. SHIRAFUJI, M. HAMADA, D. MATSUSHIGE, M. MIYAUCHI and T. MATSUURA. *J. Photochem. Photobiol., A: Chem.* **129** (1999) 121.
61. H. KOSHIMA, K. DING, T. MIURA and T. MATSUURA. *J. Photochem. Photobiol., A: Chem.* **104** (1997) 105.
62. H. KOSHIMA, D. MATSUSHIGE, M. MIYAUCHI and J. FUJITA. *Tetrahedron* **56** (2000) 6845.
63. H. KOSHIMA, K. DING, Y. CHISAKA and T. MATSUURA. *J. Am. Chem. Soc.* **118** (1996) 12059.
64. J. JACQUES, A. COLLET and S. H. WILEN. "Enantiomers, Racemates, and Resolutions", (Wiley-Interscience, New York, 1981).
65. (a) H. KOSHIMA and T. MATSUURA. *J. Synth. Org. Chem.* **56** (1998) 268. (b) H. KOSHIMA and T. MATSUURA. *J. Synth. Org. Chem.* **56** (1998) 466.
66. L. ADDADI and M. LAHAV. "Origin of Optical Activity in Nature", edited by D. C. WALKER. (Elsevier, New York, 1979) Chap. 14.
67. B. S. GREEN, M. LAHAV and D. RAVINOVICH. *Acc. Chem. Res.* **69** (1979) 191.
68. M. HASEGAWA. *Chem. Rev.* **83** (1983) 507.
69. M. SAKAMOTO. *Chem. Eur. J.* **3** (1997) 684.
70. B. L. FERINGA and R. A. VAN DELDEN. *Angew. Chem. Int. Ed.* **39** (1999) 3418.
71. For example, (a) F. H. HERBSTEIN and M. KAFTORY. *Acta Crystallogr.* **B31** (1975) 60. (b) A. THOZET and J. GAULTIER. *Acta Crystallogr.* **B31** (1977) 1051. (c) B. SHAANAN and U. SHMUELI. *Acta Crystallogr.* **B36** (1980) 2076.
72. (a) H. KOSHIMA, E. HAYASHI, T. MATSUURA, K. TANAKA and F. TODA. *Tetrahedron Lett.* **38** (1997) 15009. (b) H. KOSHIMA, E. HAYASHI and T. MATSUURA. *Supramol. Chem.* **11** (1999) 57.
73. (a) H. KOSHIMA, S. I. KHAN and M. A. GARCIA-GARIBAY. *Tetrahedron: Asymmetry* **9** (1998) 1851. (b) H. KOSHIMA and S. HONKE. *J. Org. Chem.* **64** (1999) 790. (c) H. KOSHIMA, S. HONKE and J. FUJITA. *J. Org. Chem.* **64** (1999) 3916.
74. (a) H. KOSHIMA, S. HONKE and M. MIYAUCHI. *Enantiomer* **5** (2000) 125. (b) H. KOSHIMA, M. MIYAUCHI and M. SHIRO. *Supramol. Chem.* **13** (2001) 137.

267

75. (a) S. V. EVANS, M. A. GARCIA-GARIBAY, N. OMKARAM, J. R. SCHEFFER, J. TROTTER and F. WIREKO. *J. Am. Chem. Soc.* **108** (1986) 5648. (b) M. A. GARCIA-GARIBAY, J. R. SCHEFFER, J. TROTTER and F. WIREKO. *J. Am. Chem. Soc.* **111** (1989) 4985. (c) J. CHEN, P. R. POKKULURI, J. R. SCHEFFER and J. TROTTER. *Tetrahedron Lett.* **31** (1990) 6803. (d) T. Y. FU, Z. LIU, J. R. SCHEFFER and J. TROTTER. *J. Am. Chem. Soc.* **115** (1993) 12202.

76. A. L. ROUGHTON, M. MUNEER and M. DEMUTH. *J. Am. Chem. Soc.* **115** (1993) 2085.

77. (a) F. TODA, M. YAGI and S. SODA. *J. Chem. Soc., Chem. Commun.* (1987) 1413. (b) A. SEKINE, K. HORI, Y. OHASHI, M. YAGI and F. TODA. *J. Am. Chem. Soc.* **111** (1989) 697. (c) F. TODA and H. MIYAMOTO. *J. Chem. Soc., Perkin Trans. 1* (1993) 1129. (d) D. HASHIZUME, H. KOGO, A. SEKINE, Y. OHASHI, H. MIYAMOTO and F. TODA. *J. Chem. Soc. Perkin Trans. 2* (1996) 61. (e) D. HASHIZUME, H. KOGO, A. SEKINE, Y. OHASHI, H. MIYAMOTO and F. TODA. *Acta Crystallogr.* **C51** (1995) 929.

78. (a) F. TODA and K. TANAKA. *Supramol. Chem.* **3** (1994) 87. (b) F. TODA, K. TANAKA, Z. STEIN and I. GOLDBERG. *Acta Crystallogr.* **B51** (1995) 856. (c) F. TODA, K. TANAKA, Z. STEIN and I. GOLDBERG. *Acta Crystallogr.* **C51** (1995) 2722.

79. (a) M. SAKAMOTO, M. TAKAHASHI, T. FUJITA, S. WATANABE, I. IIDA, T. NISHIO and H. AOYAMA. *J. Org. Chem.* **58** (1993) 3476. (b) M. SAKAMOTO, N. HOKARI, M. TAKAHASHI, T. FUJITA, S. WATANABE, I. IIDA and T. NISHIO. *J. Am. Chem. Soc.* **115** (1993) 818. (c) M. SAKAMOTO, M. TAKAHASHI, T. SHIMIZU, T. FUJITA, S. NISHIO, I. IKUO, K. YAMAGUCHI and S. WATANABE. *J. Org. Chem.* **60** (1995) 7088. (d) M. SAKAMOTO, M. TAKAHASHI, K. KAMIYA, K. YAMAGUCHI, T. FUJITA and S. WATANABE. *J. Am. Chem. Soc.* **118** (1996) 10664. (e) M. SAKAMOTO, M. TAKAHASHI, S. MORIIZUMI, K. YAMAGUCHI, T. FUJITA and S. WATANABE. *J. Am. Chem. Soc.* **118** (1996) 8138. (f) M. SAKAMOTO, M. TAKAHASHI, N. SEKINE, T. FUJITA, S. WATANABE and K. YAMAGUCHI. *J. Am. Chem. Soc.* **120** (1998) 12770.

80. L. ADDADI, J. MIL and M. LAHAV. *J. Am. Chem. Soc.* **104** (1982) 3422.

81. C.-M. CHUNG and M. HASEGAWA. *J. Am. Chem. Soc.* **113** (1991) 7311.

82. A. ELGAVI, B. S. GREEN and G. M. J. SCHMIDT. *J. Am. Chem. Soc.* **95** (1973) 2058.

83. T. SUZUKI, T. FUKUSHIMA, Y. YAMASHITA and T. MIYASHI. *J. Am. Chem. Soc.* **116** (1994) 2793.

84. T. SUZUKI, H. FUJII, Y. YAMASHITA, C. KABUTO, S. TANAKA, M. HARASAWA, T. MUKAI and T. MIYASHI. *J. Am. Chem. Soc.* **114** (1992) 3034.

85. H. KOSHIMA, T. NAKAGAWA, T. MATSUURA, H. MIYAMOTO and F. TODA. *J. Org. Chem.* **62** (1997) 6322.

86. E. B. FLEISCHER, N. SUNG and S. HAWKINSON. *J. Phys. Chem.* **72** (1968) 4311.

Index